PRACTICE - ASSESS - DIAG

180 Days of GEOGRAPHY for First Grade

Author
Rane Anderson

Series Consultant

Nicholas Baker, Ed.D.
Supervisor of Curriculum and Instruction
Colonial School District, DE

Publishing Credits

Corinne Burton, M.A.Ed., *Publisher*
Conni Medina, M.A.Ed., *Managing Editor*
Emily R. Smith, M.A.Ed., *Content Director*
Veronique Bos, *Creative Director*
Shaun N. Bernadou, *Art Director*
Lynette Ordoñez, *Editor*
Jodene Lynn Smith, M.A., *Editor*
Kevin Pham, *Graphic Designer*
Stephanie Bernard, *Associate Editor*

Image Credits

all images from iStock and/or Shutterstock

Standards

For information on how this resource meets national and other state standards, see pages 10–14. You may also review this information by visiting our website at www.teachercreatedmaterials.com/administrators/correlations/ and following the on-screen directions.

Shell Education
A division of Teacher Created Materials
5301 Oceanus Drive
Huntington Beach, CA 92649-1030
www.tcmpub.com/shell-education

ISBN 978-1-4258-3302-2

TABLE OF CONTENTS

INTRODUCTION

With today's geographic technology, the world seems smaller than ever. Satellites can accurately measure the distance between any two points on the planet and give detailed instructions about how to get there in real time. This may lead some people to wonder why we still study geography.

While technology is helpful, it isn't always accurate. We may need to find detours around construction, use a trail map, outsmart our technology, and even be the creators of the next navigational technology.

But geography is also the study of cultures and how people interact with the physical world. People change the environment, and the environment affects how people live. People divide the land for a variety of reasons. Yet no matter how it is divided or why, people are at the heart of these decisions. To be responsible and civically engaged, students must learn to think in geographical terms.

The Need for Practice

To be successful in geography, students must understand how the physical world affects humanity. They must not only master map skills but also learn how to look at the world through a geographical lens. Through repeated practice, students will learn how a variety of factors affect the world in which they live.

Understanding Assessment

In addition to providing opportunities for frequent practice, teachers must be able to assess students' geographical understandings. This allows teachers to adequately address students' misconceptions, build on their current understandings, and challenge them appropriately. Assessment is a long-term process that involves careful analysis of student responses from a discussion, project, practice sheet, or test. The data gathered from assessments should be used to inform instruction: slow down, speed up, or reteach. This type of assessment is called *formative assessment.*

HOW TO USE THIS BOOK

Weekly Structure

The first two weeks of the book focus on map skills. By introducing these skills early in the year, students will have a strong foundation on which to build throughout the year. Each of the remaining 34 weeks will follow a regular weekly structure.

Each week, students will study a grade-level geography topic and a location in a community. Locations may be a town, a street, a home, or a school.

Days 1 and 2 of each week focus on map skills. Days 3 and 4 allow students to apply information and data to what they have learned. Day 5 helps students connect what they have learned to themselves.

Day 1—Reading Maps: Students will study a grade-appropriate map and answer questions about it.

Day 2—Creating Maps: Students will create maps or add to an existing map.

Day 3—Read About It: Students will read a text related to the topic or location for the week and answer text-dependent or photo-dependent questions about it.

Day 4—Think About It: Students will analyze a chart, diagram, or other graphic related to the topic or location for the week and answer questions about it.

Day 5—Geography and Me: Students will do an activity to connect what they learned to themselves.

Five Themes of Geography

Good geography teaching encompasses all five themes of geography: location, place, human-environment interaction, movement, and region. Location refers to the absolute and relative locations of a specific point or place. The place theme refers to the physical and human characteristics of a place. Human-environment interaction describes how humans affect their surroundings and how the environment affects the people who live there. Movement describes how and why people, goods, and ideas move between different places. The region theme examines how places are grouped into different regions. Regions can be divided based on a variety of factors, including physical characteristics, cultures, weather, and political factors.

HOW TO USE THIS BOOK *(cont.)*

Weekly Themes

The following chart shows the topics, states, and themes of geography that are covered during each week of instruction.

Wk.	Topic	Location	Geography Themes
1	—Map Skills Only—		Location
2			Location
3	Reading a Map	Backyard	Location
4	Community	Town	Place
5	Near vs. Far	Neighborhood	Location, Movement
6	Addresses	Street	Location
7	Routes	Neighborhood	Location
8	Regions and Climate	Poles and Equator	Place, Region
9	Rural Communities	Rural Community	Place
10	Landforms	Wilderness	Place
11	Renewable Resource	Wilderness	Human-Environment Interaction
12	Bodies of Water	Lake Baikal	Place
13	Canyons	Grand Canyon	Place
14	Renewable Resource	Wilderness	Human-Environment Interaction
15	Urban Communities	Shanghai, China	Place, Human-Environment Interaction
16	Non-renewable Resource	Wilderness	Human-Environment Interaction
17	Transportation	Venice, Italy	Place, Human-Environment Interaction, Movement
18	Landmarks	City	Location

HOW TO USE THIS BOOK *(cont.)*

Wk.	Topic	Location	Geography Themes
19	Oil Spills	Coast	Human-Environment Interaction
20	Population	Australia	Region
21	Transportation	City	Movement
22	Rivers	Nile River	Human-Environment Interaction, Movement
23	Suburban Communities	Suburban Community	Location, Place
24	Farmer's Market	City	Location, Movement
25	School	Classroom	Location
26	Parks and Gardens	Backyard	Place
27	Tourism	Savanna	Human-Environment Interaction, Region
28	Climate	Russia	Place, Human-Environment Interaction
29	Islands	Hawai'ian Islands	Place
30	Going to School	School	Location, Place
31	Geocaches	Playground	Location
32	Mountains	Mount Everest	Place, Human-Environment Interaction
33	Routes	Hiking Trail	Human-Environment Interaction
34	Volcanoes	Mount St. Helens	Location, Place
35	Cliffs	Shoreline	Place, Human-Environment Interaction
36	Food Around the World	City	Movement

HOW TO USE THIS BOOK *(cont.)*

Using the Practice Pages

The activity pages provide practice and assessment opportunities for each day of the school year. Teachers may wish to prepare packets of weekly practice pages for the classroom or for homework.

As outlined on page 4, each week examines one location and one geography topic.

The first two days focus on map skills. On Day 1, students will study a map and answer questions about it. On Day 2, they will add to or create a map.

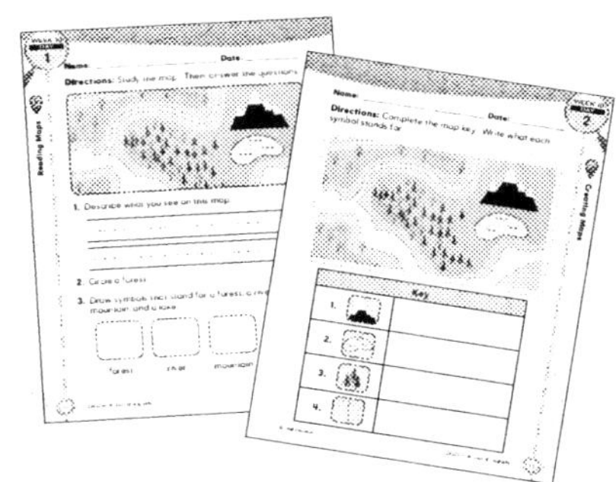

Days 3 and 4 allow students to apply information and data from texts, charts, graphs, and other sources to the location being studied.

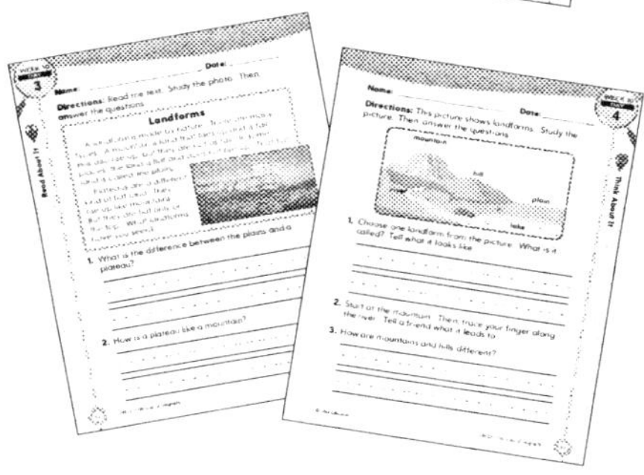

On Day 5, students will apply what they learned to themselves.

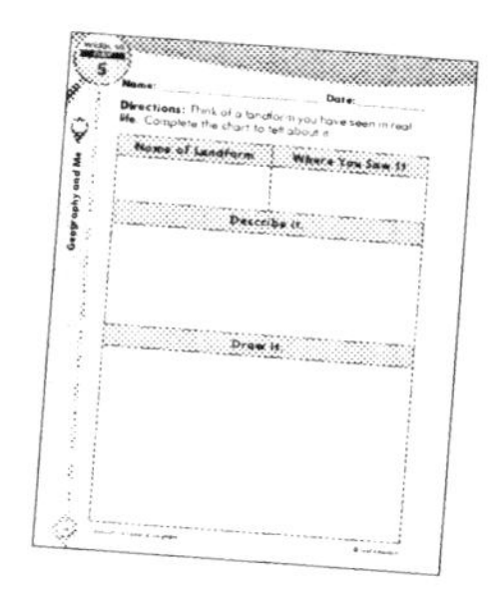

Using the Resources

Rubrics for the types of days (map skills, applying information and data, and making connections) can be found on pages 202–204 and in the Digital Resources. Use the rubrics to assess students' work. Be sure to share these rubrics with students often so that they know what is expected of them.

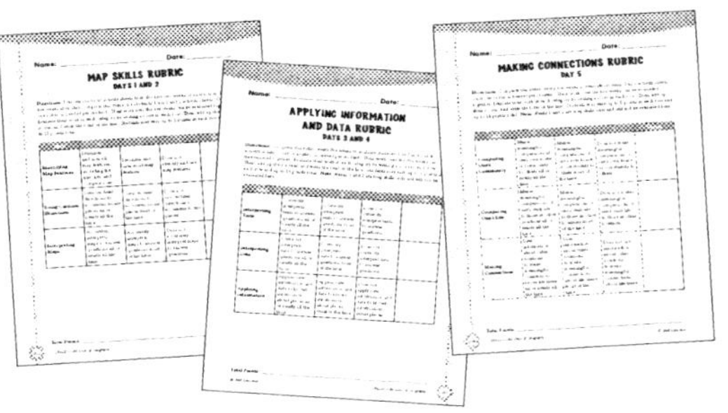

HOW TO USE THIS BOOK *(cont.)*

Diagnostic Assessment

Teachers can use the practice pages as diagnostic assessments. The data analysis tools included with the book enable teachers or parents to quickly score students' work and monitor their progress. Teachers and parents can quickly see which skills students may need to target further to develop proficiency.

Students will learn map skills, how to apply text and data to what they have learned, and how to relate what they learned to themselves. You can assess students' learning in each area using the rubrics on pages 202–204. Then, record their scores on the Practice Page Item Analysis sheets on pages 205–207. These charts are also provided in the Digital Resources as PDFs, Microsoft Word® files, and Microsoft Excel® files (filenames: skillsanalysis.pdf, skillsanalysis.docx, skillsanalysis.xlsx; dataanalysis.pdf, dataanalysis.docx, dataanalysis.xlsx; connectanalysis.pdf, connectanalysis.docx, connectanalysis.xlsx). Teachers can input data into the electronic files directly on the computer, or they can print the pages.

To Complete the Practice Page Item Analyses:

- Write or type students' names in the far-left column. Depending on the number of students, more than one copy of the forms may be needed.
    - The skills are indicated across the tops of the pages.
    - The weeks in which students should be assessed are indicated in the first rows of the charts. Students should be assessed at the ends of those weeks.
- Review students' work for the days indicated in the chart. For example, if using the Making Connections Analysis sheet for the first time, review students' work from Day 5 for all five weeks.
- Add the scores for each student. Place that sum in the far right column. Record the class average in the last row. Use these scores as benchmarks to determine how students are performing.

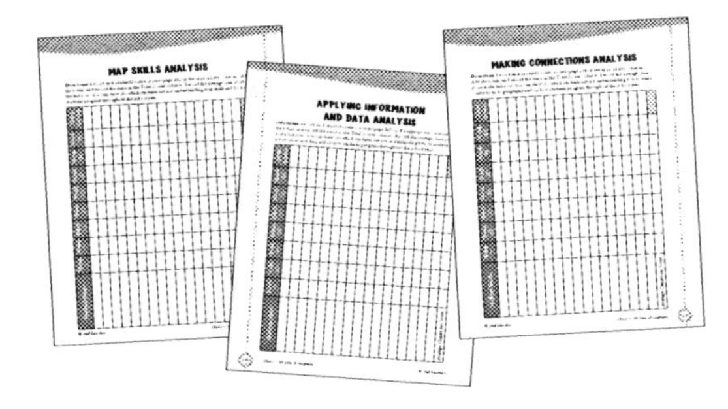

Digital Resources

The Digital Resources contain digital copies of the rubrics, analysis pages, and standards charts. See page 208 for more information.

HOW TO USE THIS BOOK *(cont.)*

Using the Results to Differentiate Instruction

Once results are gathered and analyzed, teachers can use them to inform the way they differentiate instruction. The data can help determine which geography skills are the most difficult for students and which students need additional instructional support and continued practice.

Whole-Class Support

The results of the diagnostic analysis may show that the entire class is struggling with certain geography skills. If these concepts have been taught in the past, this indicates that further instruction or reteaching is necessary. If these concepts have not been taught in the past, this data is a great preassessment and may demonstrate that students do not have a working knowledge of the concepts. Thus, careful planning for the length of the unit(s) or lesson(s) must be considered, and additional front-loading may be required.

Small-Group or Individual Support

The results of the diagnostic analysis may show that an individual student or a small group of students is struggling with certain geography skills. If these concepts have been taught in the past, this indicates that further instruction or reteaching is necessary. Consider pulling these students aside to instruct them further on the concepts while others are working independently. Students may also benefit from extra practice using games or computer-based resources.

Teachers can also use the results to help identify proficient individual students or groups of students who are ready for enrichment or above-grade-level instruction. These students may benefit from independent learning contracts or more challenging activities.

STANDARDS CORRELATIONS

Shell Education is committed to producing educational materials that are research and standards based. In this effort, we have correlated all our products to the academic standards of all 50 states, the District of Columbia, the Department of Defense Dependents Schools, and all Canadian provinces.

How to Find Standards Correlations

To print a customized correlation report of this product for your state, visit our website at **www.teachercreatedmaterials.com/administrators/correlations** and follow the on-screen directions. If you require assistance in printing correlation reports, please contact our Customer Service Department at 1-877-777-3450.

Purpose and Intent of Standards

The Every Student Succeeds Act (ESSA) mandates that all states adopt challenging academic standards that help students meet the goal of college and career readiness. While many states already adopted academic standards prior to ESSA, the act continues to hold states accountable for detailed and comprehensive standards. Standards are designed to focus instruction and guide adoption of curricula. Standards are statements that describe the criteria necessary for students to meet specific academic goals. They define the knowledge, skills, and content students should acquire at each level. Standards are also used to develop standardized tests to evaluate students' academic progress. Teachers are required to demonstrate how their lessons meet state standards. State standards are used in the development of our products, so educators can be assured they meet the academic requirements of each state.

The activities in this book are aligned to the National Geography Standards and the McREL standards. The chart on pages 11–12 lists the National Geography Standards used throughout this book. The chart on pages 13–14 correlates the specific McREL and National Geography Standards to each week. The standards charts are also in the Digital Resources (standards.pdf).

C3 Framework

This book also correlates to the College, Career, and Civic Life (C3) Framework published by the National Council for the Social Studies. By completing the activities in this book, students will learn to answer and develop strong questions (Dimension 1), critically think like a geographer (Dimension 2), and effectively choose and use geography resources (Dimension 3). Many activities also encourage students to take informed action within their communities (Dimension 4).

STANDARDS CORRELATIONS *(cont.)*

180 Days of Geography is designed to give students daily practice in geography through engaging activities. Students will learn map skills, how to apply information and data to their understandings of various locations and cultures, and how to apply what they learned to themselves.

Easy to Use and Standards Based

There are 18 National Geography Standards, which fall under six essential elements. Specific expectations are given for fourth grade, eighth grade, and twelfth grade. For this book, fourth grade expectations were used with the understanding that full mastery is not expected until that grade level.

Essential Elements	National Geography Standards
The World in Spatial Terms	**Standard 1:** How to use maps and other geographic representations, geospatial technologies, and spatial thinking to understand and communicate information
	Standard 2: How to use mental maps to organize information about people, places, and environments in a spatial context
	Standard 3: How to analyze the spatial organization of people, places, and environments on Earth's surface
Places and Regions	**Standard 4:** The physical and human characteristics of places
	Standard 5: People create regions to interpret Earth's complexity
	Standard 6: How culture and experience influence people's perceptions of places and regions
Physical Systems	**Standard 7:** The physical processes that shape the patterns of Earth's surface
	Standard 8: The characteristics and spatial distribution of ecosystems and biomes on Earth's surface

STANDARDS CORRELATIONS *(cont.)*

Essential Elements	National Geography Standards
Human Systems	**Standard 9:** The characteristics, distribution, and migration of human populations on Earth's surface
	Standard 10: The characteristics, distribution, and complexity of Earth's cultural mosaics
	Standard 11: The patterns and networks of economic interdependence on Earth's surface
	Standard 12: The process, patterns, and functions of human settlement
	Standard 13: How the forces of cooperation and conflict among people influence the division and control of Earth's surface
Environment and Society	**Standard 14:** How human actions modify the physical environment
	Standard 15: How physical systems affect human systems
	Standard 16: The changes that occur in the meaning, use, distribution, and importance of resources
The Uses of Geography	**Standard 17:** How to apply geography to interpret the past
	Standard 18: How to apply geography to interpret the present and plan for the future

–2012 National Council for Geographic Education

STANDARDS CORRELATIONS *(cont.)*

Easy to Use and Standards Based *(cont.)*

This chart lists the specific National Geography Standards and McREL standards that are covered each week.

Wk.	NGS	McREL Standards
1	Standards 1 and 3	Identifies physical and human features in terms of the four spatial elements. Knows the absolute and relative location of a community and places within it.
2	Standards 1 and 3	Identifies physical and human features in terms of the four spatial elements. Knows the absolute and relative location of a community and places within it.
3	Standards 1 and 3	Identifies physical and human features in terms of the four spatial elements.
4	Standard 4	Knows the physical and human characteristics of the local community.
5	Standard 12	Knows the absolute and relative location of a community and places within it.
6	Standard 1	Knows the absolute and relative location of a community and places within it.
7	Standard 1	Knows the absolute and relative location of a community and places within it.
8	Standard 5	Knows areas that can be classified as regions according to physical criteria.
9	Standard 12	Knows that places can be defined in terms of their predominant human and physical characteristics.
10	Standard 7	Knows that places can be defined in terms of their predominant human and physical characteristics.
11	Standard 16	Knows ways in which people depend on the physical environment. Knows the role that resources play in our daily lives.
12	Standard 7	Knows that places can be defined in terms of their predominant human and physical characteristics.
13	Standard 7	Knows that places can be defined in terms of their predominant human and physical characteristics.
14	Standard 16	Knows ways in which people depend on the physical environment. Knows the role that resources play in our daily lives.
15	Standard 12	Knows the similarities and differences in housing and land use in urban and suburban areas.
16	Standard 16	Knows the role that resources play in our daily lives.
17	Standard 11	Knows the modes of transportation used to move people, products and ideas from place to place, their importance, and their advantages and disadvantages.
18	Standard 4	Knows the absolute and relative location of a community and places within it.

STANDARDS CORRELATIONS *(cont.)*

Wk.	NGS	McREL Standards
19	Standard 14	Knows ways that people solve common problems by cooperating.
20	Standards 5 and 12	Knows areas that can be classified as regions according to physical criteria and human criteria.
21	Standard 11	Knows the modes of transportation used to move people, products and ideas from place to place, their importance, and their advantages and disadvantages.
22	Standard 15	Knows ways in which people depend on the physical environment.
23	Standard 12	Knows the absolute and relative location of a community and places within it. Knows the similarities and differences in housing and land use in urban and suburban areas.
24	Standard 11	Knows the absolute and relative location of a community and places within it.
25	Standard 10	Knows the basic components of culture.
26	Standard 4	Knows that places can be defined in terms of their predominant human and physical characteristics.
27	Standards 5 and 14	Knows areas that can be classified as regions according to physical criteria and human criteria. Knows how areas of a community have changed over time.
28	Standard 15	Understands the globe as a representation of the Earth. Knows that places can be defined in terms of their predominant human and physical characteristics.
29	Standard 7	Knows that places can be defined in terms of their predominant human and physical characteristics.
30	Standard 10	Knows the basic components of culture.
31	Standard 3	Knows the absolute and relative location of a community and places within it.
32	Standard 7	Knows that places can be defined in terms of their predominant human and physical characteristics.
33	Standard 3	Identifies physical and human features in terms of the four spatial elements.
34	Standard 7	Knows that places can be defined in terms of their predominant human and physical characteristics.
35	Standards 7 and 15	Knows that places can be defined in terms of their predominant human and physical characteristics. Knows ways that people solve common problems by cooperating.
36	Standard 10	Knows the basic components of culture.

Name: ______________________ **Date:** ____________

Directions: Study the map. Then, answer the questions. Use the Word Bank to help you.

My Bedroom

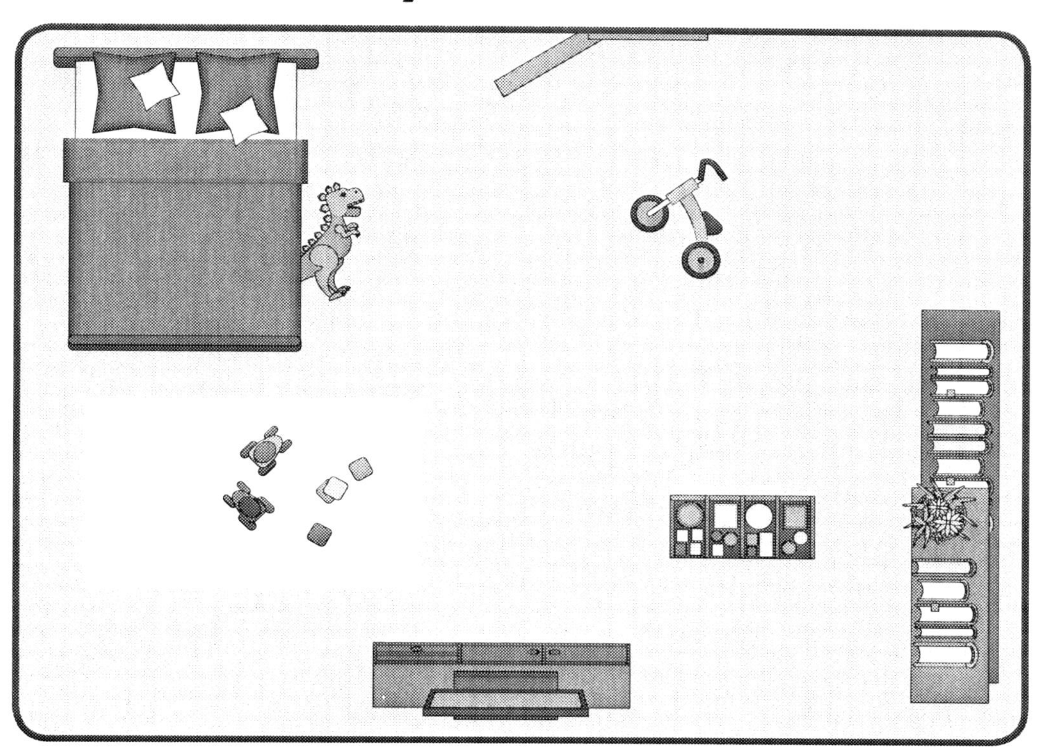

Word Bank		
next to	under	between

1. What is the title of this map?

____

____

2. The dinosaur is ______________ the bed.

3. Draw a car on top of the rug.

Name: ______________________ **Date:** ____________

Directions: A symbol is a picture that stands for something. Draw lines to match the pictures of the real objects to the symbols.

Real Objects	Symbols
1.	
2.	
3.	
4.	
5.	
6.	

Name: ______________________________ **Date:** ____________

Directions: On a map, a symbol will look small and simple. Study the map. Then, answer the questions.

Map Symbols

1. Circle the symbol for the park on the map.

2. Draw the house symbol in three more places on the map.

3. Draw a box around the symbol for the zoo.

4. Create a symbol for an ice cream store. Draw it on the map.

Name: ______________________________ **Date:** ______________

Directions: A compass rose shows the cardinal directions. Use north, south, east, or west to answer the questions.

North
West East
South

1. Name one thing you see on the east side of the picture.

____

2. Which side of the picture is the slide mostly found in? Circle your answer.

north south east west

3. "**N**ever **E**at **S**limy **W**orms" can help you keep track of the order of the cardinal directions. Tell a friend a different way to remember them.

Name: ________________________ **Date:** ____________

Directions: Study the map. Answer the questions.

Treasure Map

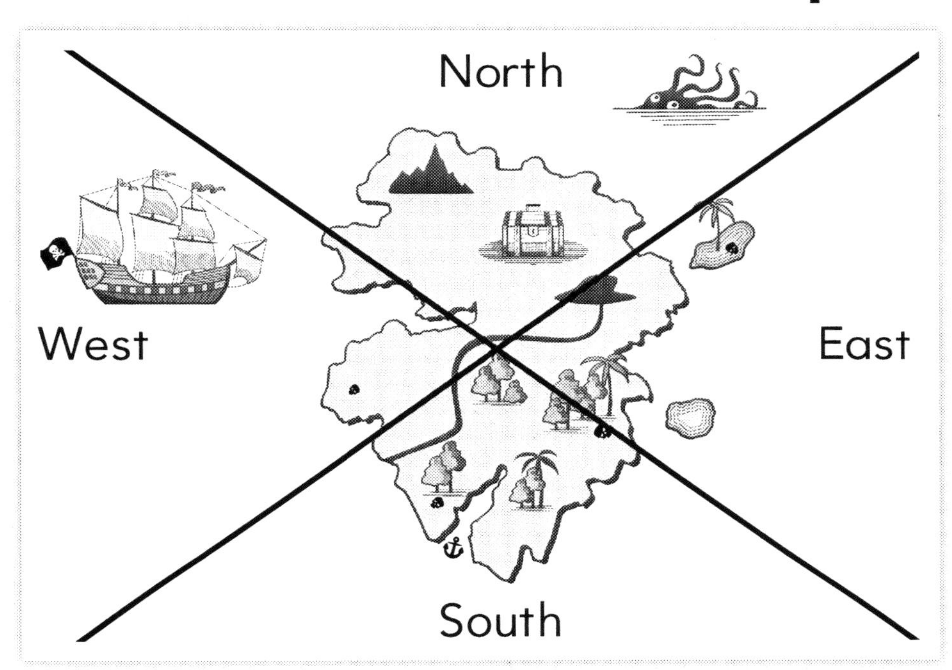

1. Label the compass rose with the cardinal directions. (Hint: Never Eat Slimy Worms)

2. Name one thing you see in the northern part of the map.

____

3. The forest is ____________________ of the river.

4. The ship is ____________________ of the island.

Map Skills

Name: ______________________ **Date:** ____________

Directions: Label the compass rose with the cardinal directions. Then, complete the story.

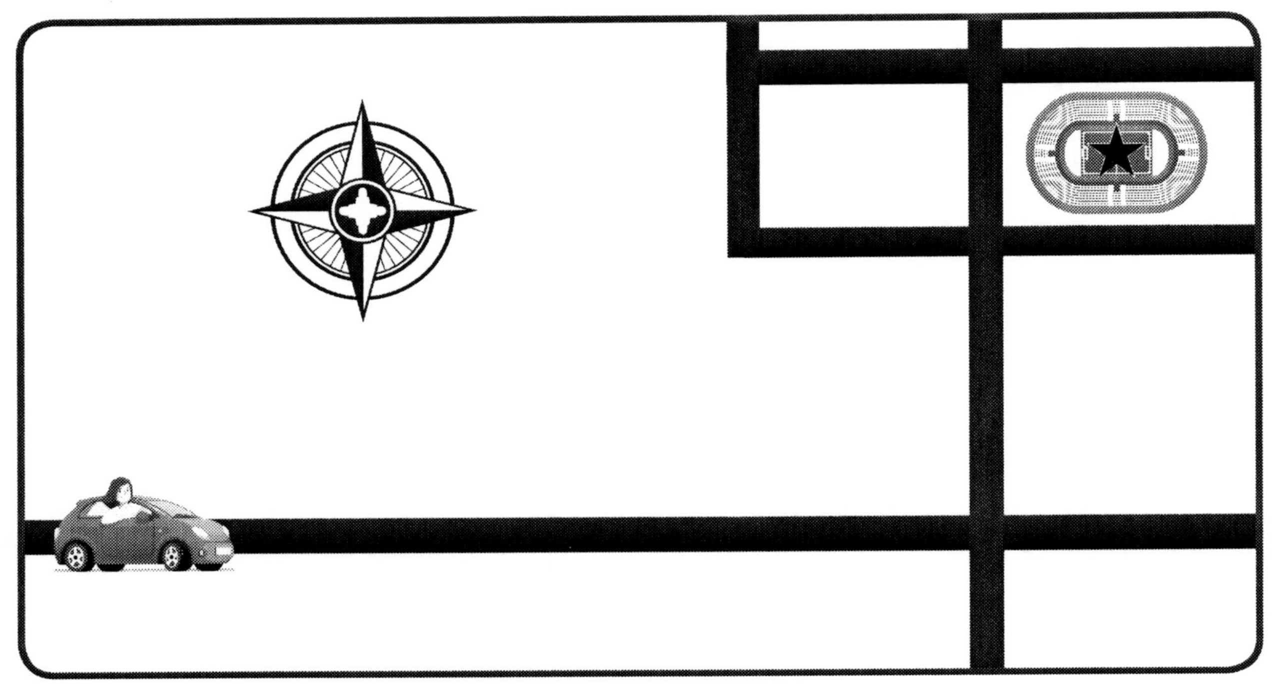

Word Bank		
north	east	right

A driver needs to get to the stadium.

She starts by heading ______________________.

Then, she turns left.

Now, she is heading ______________________.

Then, she turns ______________________ into the stadium.

Name: ______________________ **Date:** ____________

Directions: Street names help people find their way around town. Study the map. Then, answer the questions.

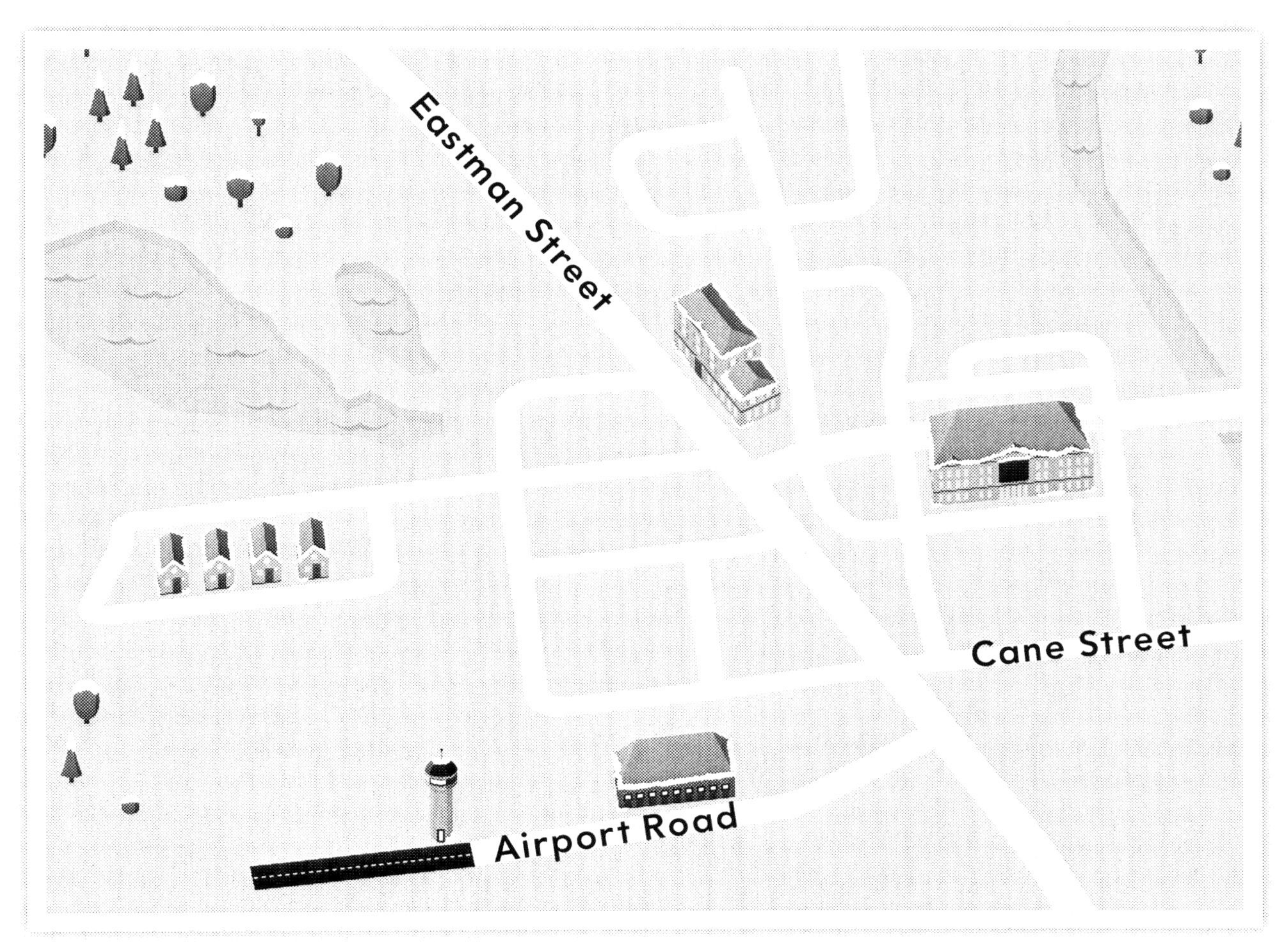

1. Draw a park near Cane Street.

2. Draw an airplane near Airport Road.

3. Draw a house on Eastman Street.

4. What is the name of the main road that crosses through the town?

____

Name: ______________________ **Date:** ____________

Directions: Study the street names on this map. Then, answer the questions.

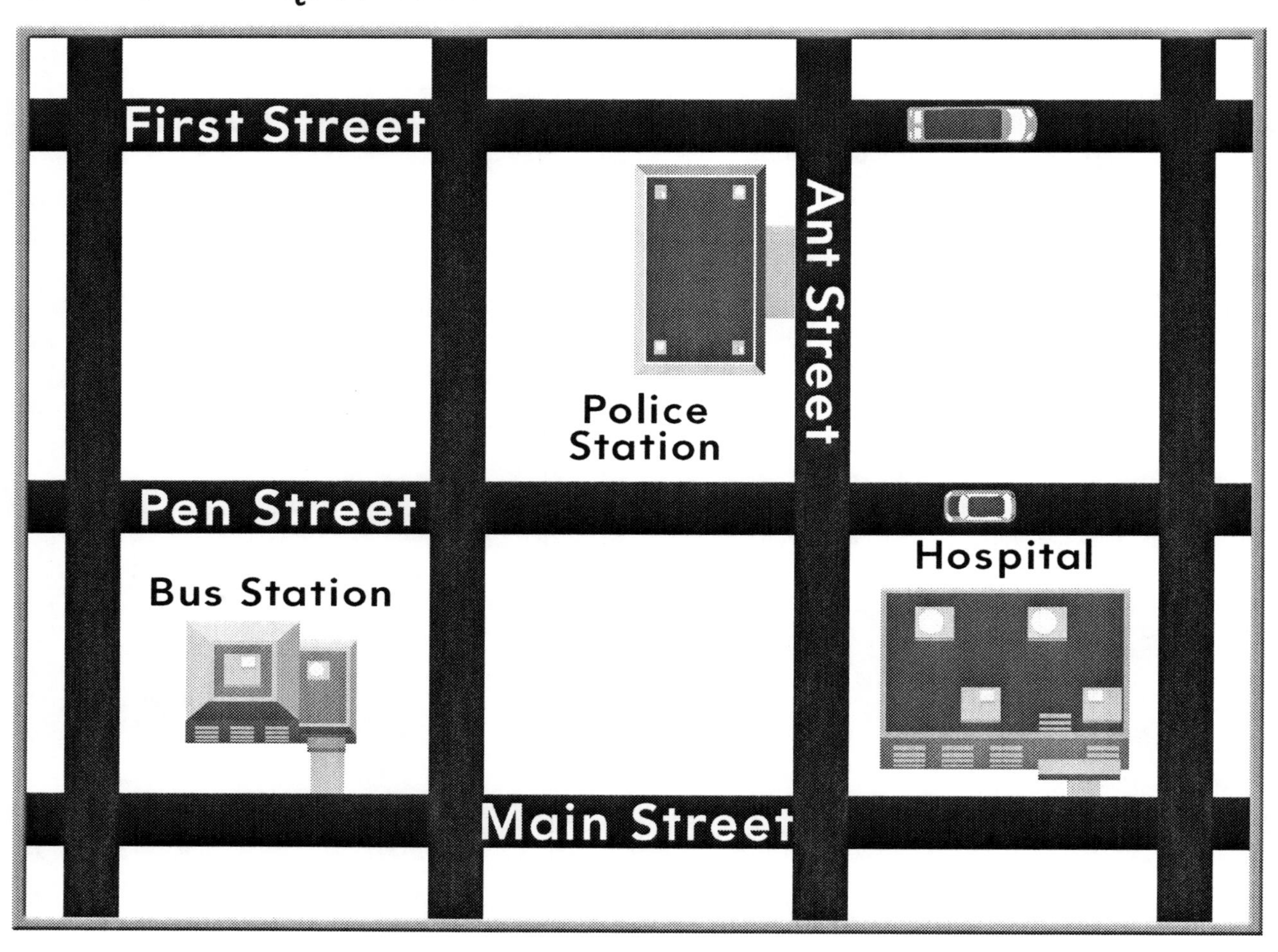

1. The police station is on ______________ Street.

2. The hospital is on ______________ Street.

3. What street is the school bus driving down?

____

Name: ______________________ Date: ____________

Directions: Study the map. Write names for the different places on this map.

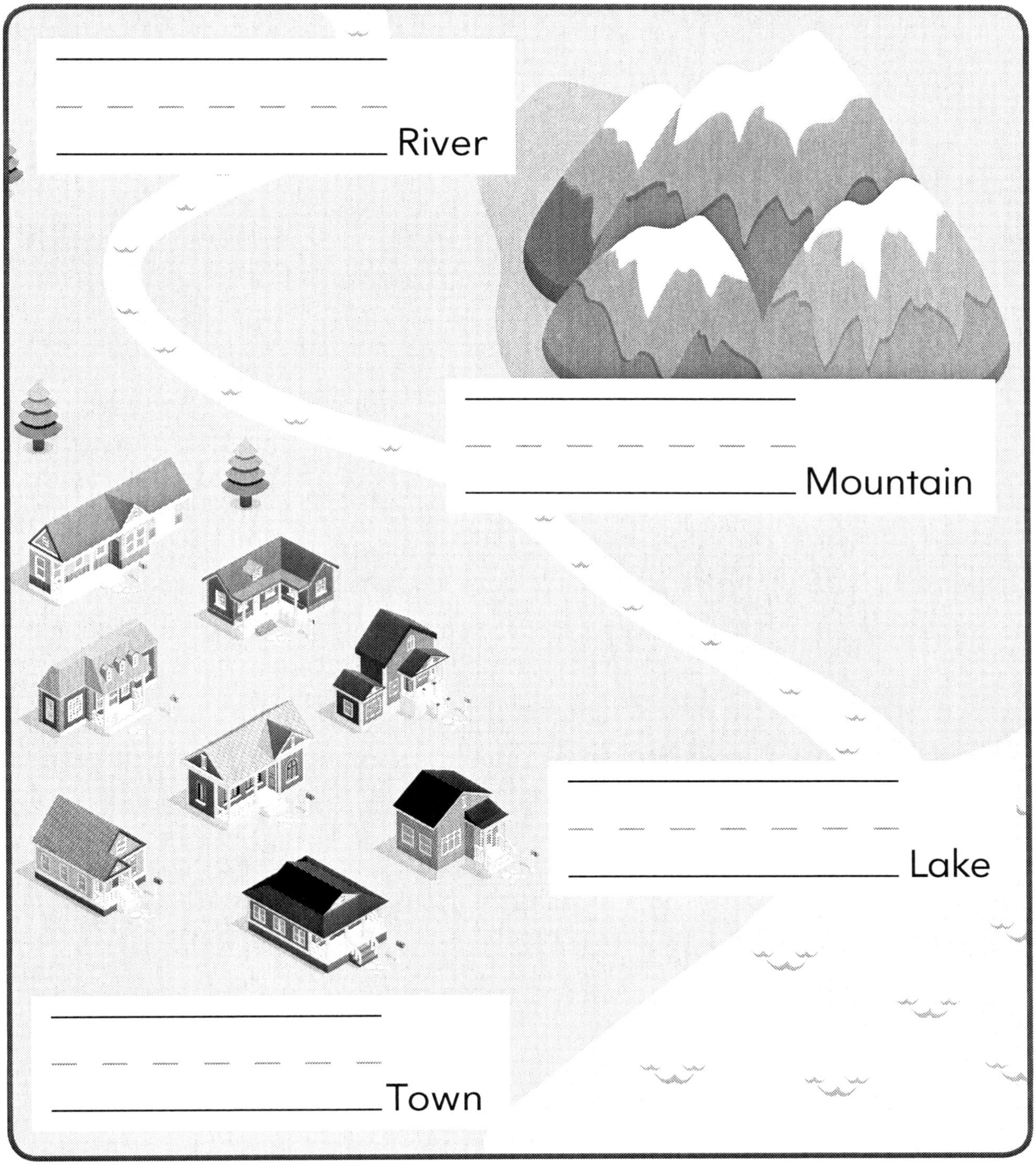

______________ River

______________ Mountain

______________ Lake

______________ Town

Name: ______________________ **Date:** ____________

Directions: Study the map. Then, follow the steps.

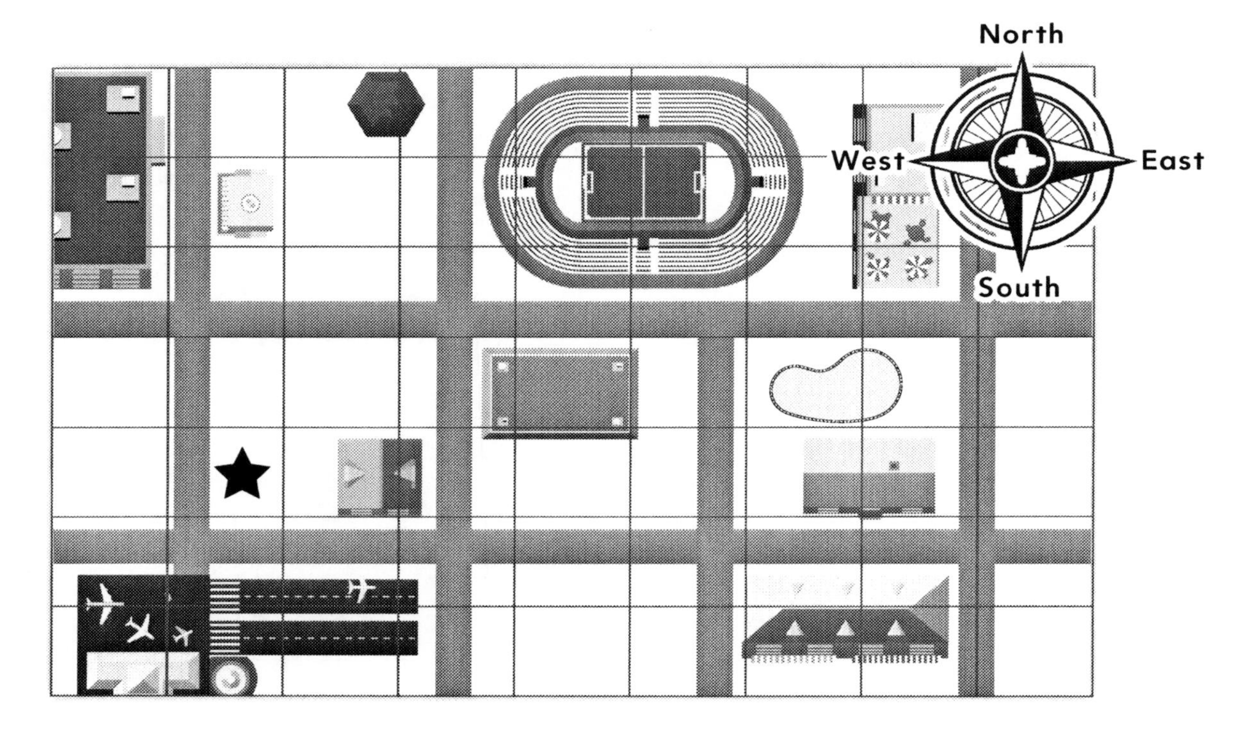

1. Start at the star. Move 3 squares north. Color the building in that square.

2. Now, move 3 squares east. What is inside of that square?

____

3. Move 5 squares south and 2 squares east. Color the building in that square.

4. Which direction would you have to move to get to the pond?

____

Name: ______________________ **Date:** ____________

Directions: Study the map of the house and backyard. Then, answer the questions.

1. Name two things in the backyard.

______________________ ______________________

2. Where would be the best place to put swings?

____

____

Creating Maps

Name: ______________________________ **Date:** ______________

Directions: Follow the steps.

1. Draw a flower next to the swimming pool.
2. Circle a place to sit.
3. Trace the path to the house.

Name: ______________________ **Date:** ____________

Directions: Read the text. Study the photo. Then, answer the questions.

Aerial Maps

From the ground, buildings look big and streets look wide. But have you looked out of an airplane window? You can see the whole city from the sky. Looking at the ground from high up is called an *aerial view*. From up there, buildings and houses look like tiny boxes. A map is a drawing that shows a part of the world. A map looks like it was drawn from high up.

1. What is an aerial view?

____

____

2. What do houses look like from up high?

____

____

WEEK 3
DAY
4

Think About It

Name: ______________________ Date: ____________

Directions: This is an aerial view of a football stadium. Study the picture. Then, answer the questions.

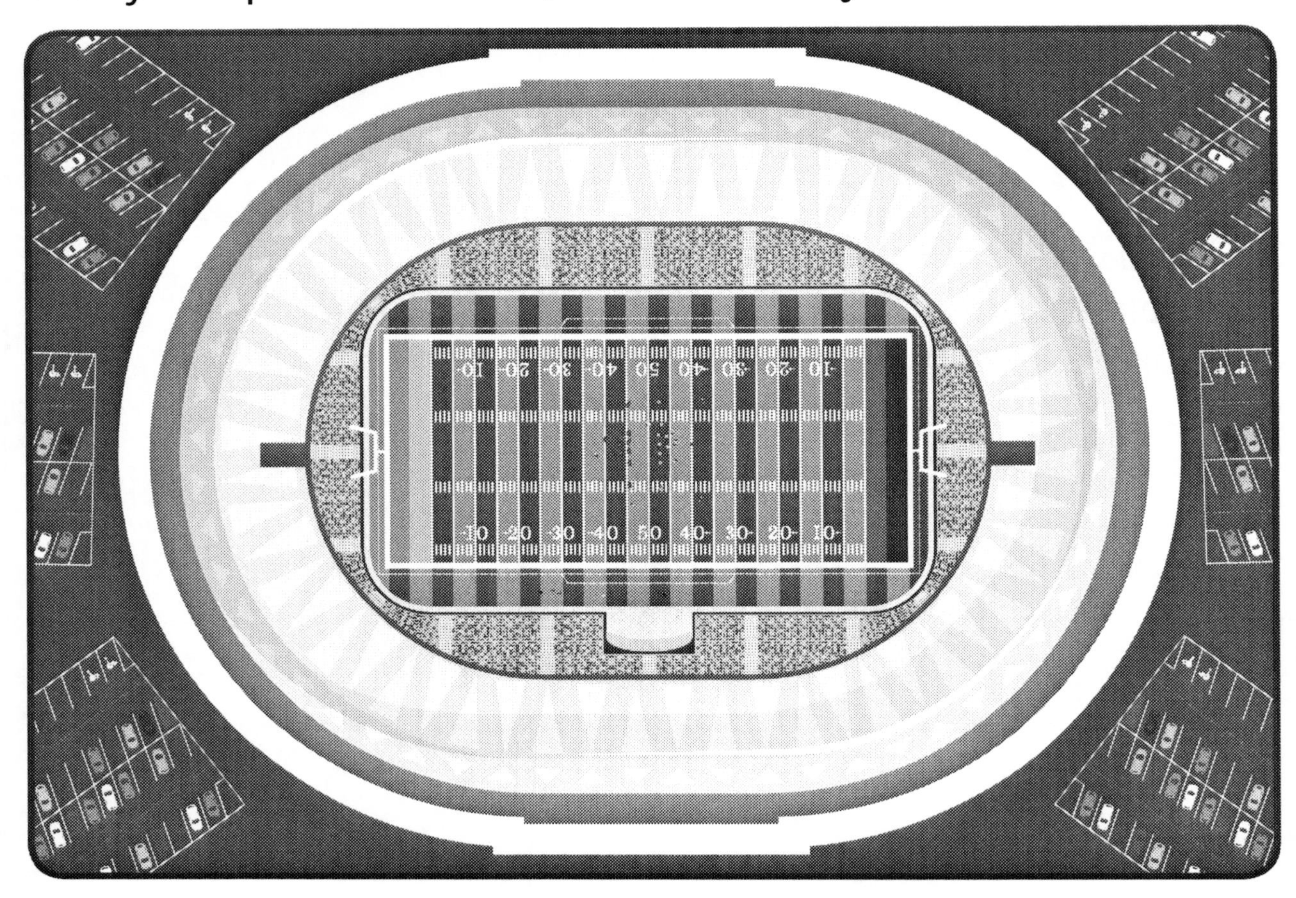

1. Circle the goalposts.

2. Draw boxes where the fans sit.

3. Buses brought people to the game. Describe where cars might be parked.

____

____

Name: ______________________ Date: ____________

Directions: Draw and write about what an aerial view of your home looks like.

Name: ______________________ **Date:** ____________

Directions: Study the map. Then, answer the questions.

1. Name places where kids like to play.

______________________ ______________________

2. Where could someone buy dessert in this community?

____

3. How many gas stations are in this community?

____

Name: ______________________________ **Date:** ______________

Directions: Draw something from your community that is not on the map. Draw it in the box.

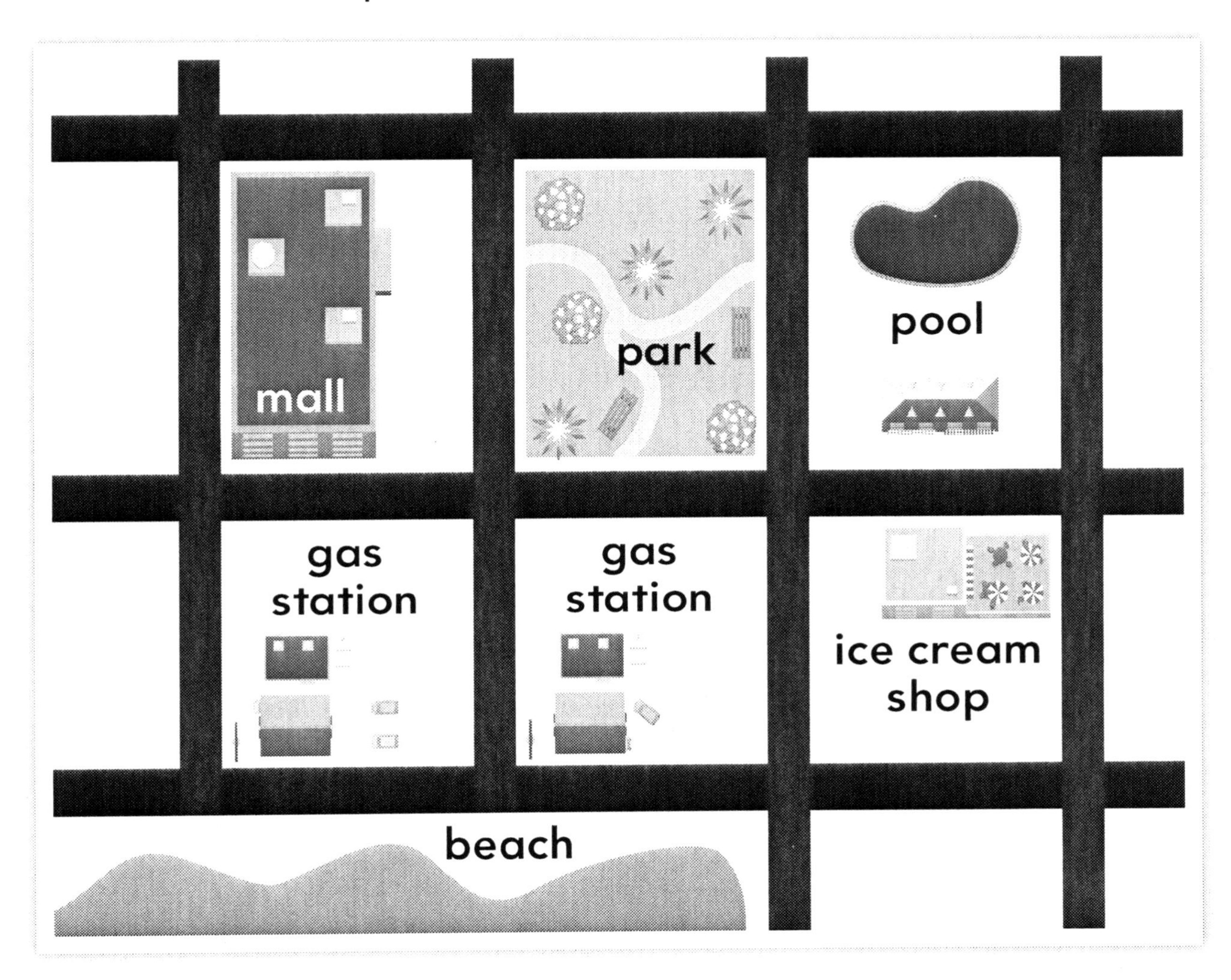

Name: ______________________ **Date:** ______________

Read About It

Directions: Read the text. Study the photo. Then, answer the questions.

A Community

What is a community? It is a place where people live. It can be a city or a small town. People work there. They shop there. And they play there. There are special places in a community. There might be a zoo. There might be an airport. A community is made up of places to go, things to do, and people to see.

1. Name two places people go in a community.

______________________ ______________________

2. Is a museum part of a community? How do you know?

____

____

Think About It

Name: ______________________ **Date:** ____________

Directions: A farmers' market is a special place in a community. Study the photo. Answer the questions.

1. Write about what happens at a farmers' market.

____

____

2. List things people can buy at a farmers' market.

____________________ ____________________

____________________ ____________________

Geography and Me

Name: ______________________ **Date:** ____________

Directions: Draw and write about a special place in your community.

Name: ______________________ **Date:** ____________

Directions: Study the map. Then, answer the questions.

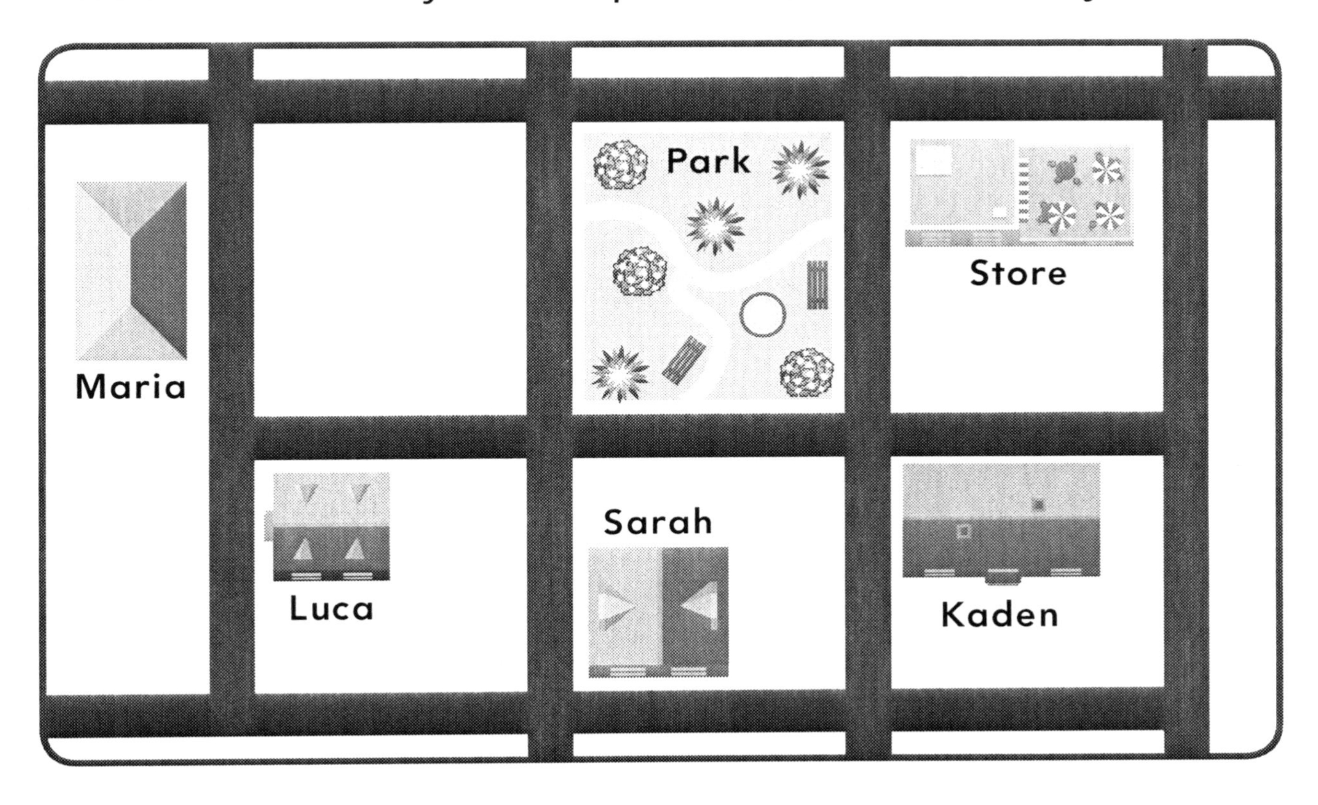

1. Is the park farther from Kaden's house or Maria's house?

____

2. Would it take Maria longer to walk to Luca's house or the store? Why?

____

____

____

Name: ______________________________ **Date:** ______________

Directions: Sarah and Maria are playing hide-and-seek at the park. Draw Sarah hiding far from a bench. Draw Maria near a bush.

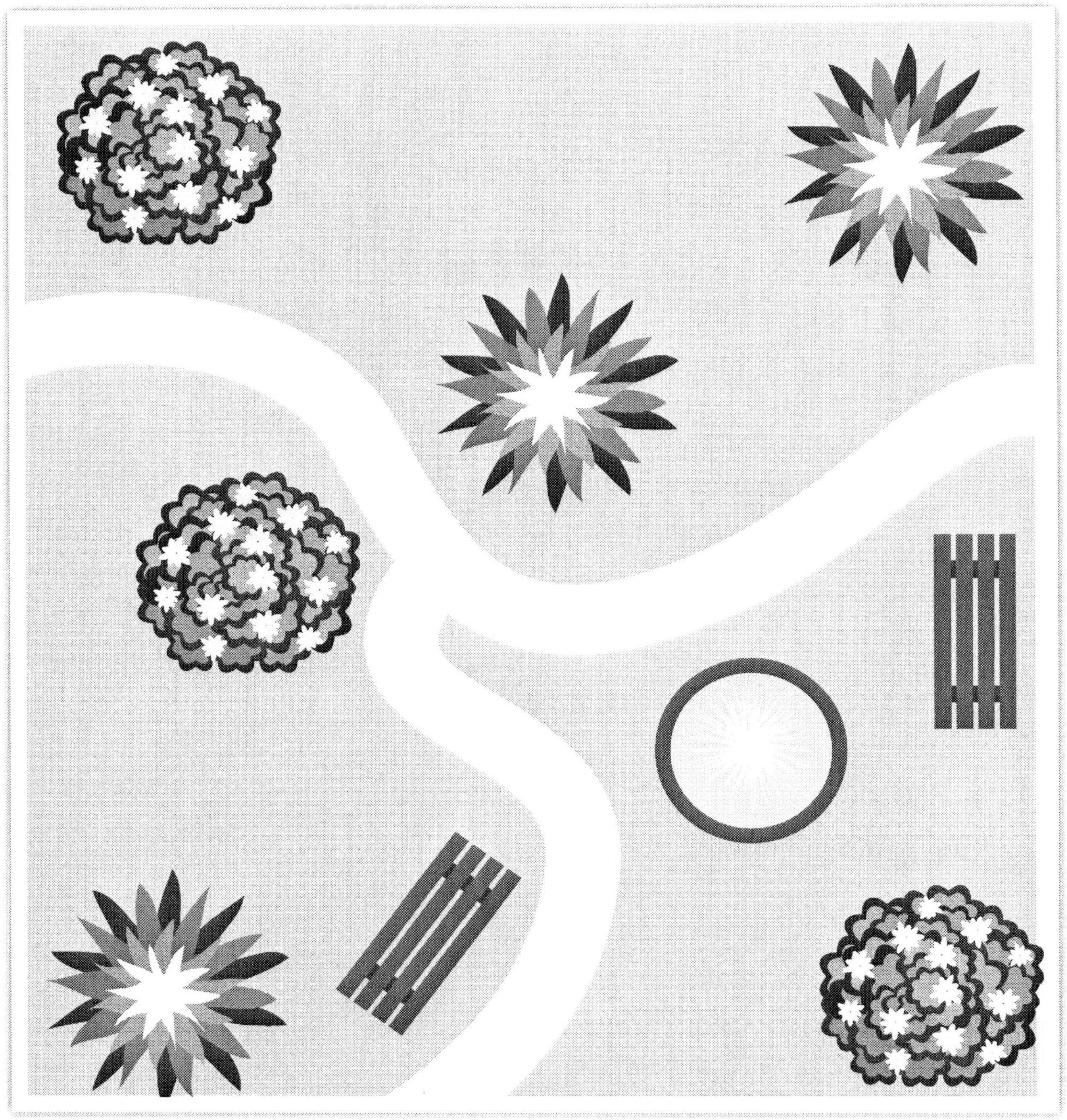

Name: ______________________ **Date:** ____________

Directions: Read the text. Study the photo. Then, answer the questions.

Moving Near and Far

People move to new places from near and far. Some people move just 10 minutes away. Other people move to different countries. They are immigrants. They leave to make a home in a new country. They may move to stay safe or to find new jobs. When they move, they find new houses. Adults get new jobs. Kids go to new schools. And they all make new friends.

1. Why do some people move to a new country?

____

____

2. What do people do when they get to a new country?

____

____

Think About It

Name: ______________________ **Date:** ____________

Directions: Sam lives in New York City. Three new kids moved onto his block. The chart shows the distance each kid moved. Use the chart to answer the questions.

	Rosie	Jaden	Marcus
Distance Moved	5 miles	15 miles	1,000 miles
Place	Queens	Bronx	Canada

1. Who moved the farthest?

2. Marcus is an immigrant. What country did he move from?

3. How do you think Marcus feels after moving?

Name: ______________________________ Date: ______________

Directions: Draw and write about a place you went that was far from your home. Include where you went and how you got there.

Name: ______________________ **Date:** ____________

Directions: Study the map. Then, answer the questions.

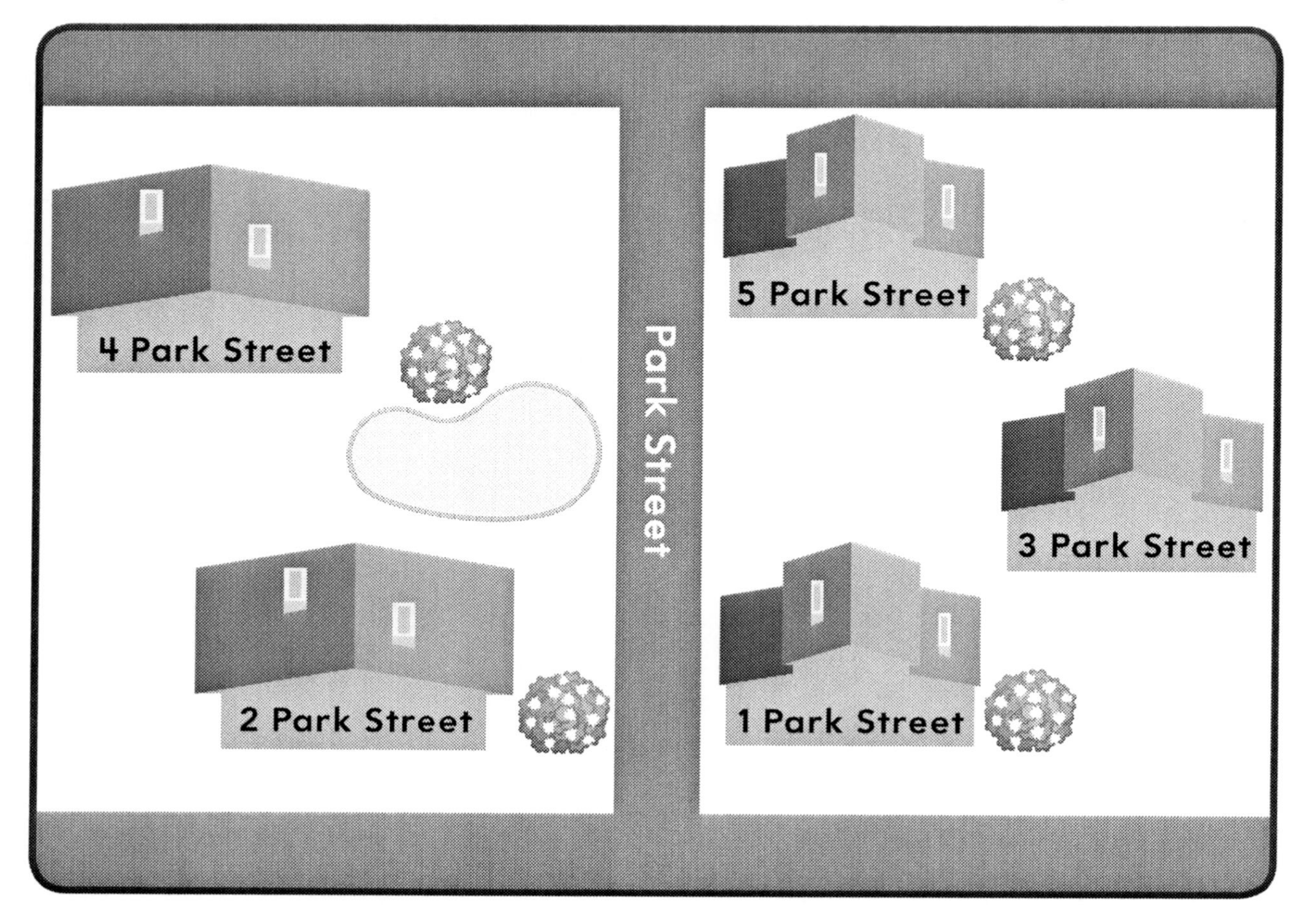

1. What is the name of this street?

____

2. What pattern do you see in the house numbers?

____

____

Name: ______________________________ **Date:** ______________

Directions: Draw a line from each envelope to the matching house. Draw a route a mail carrier might take to deliver these envelopes.

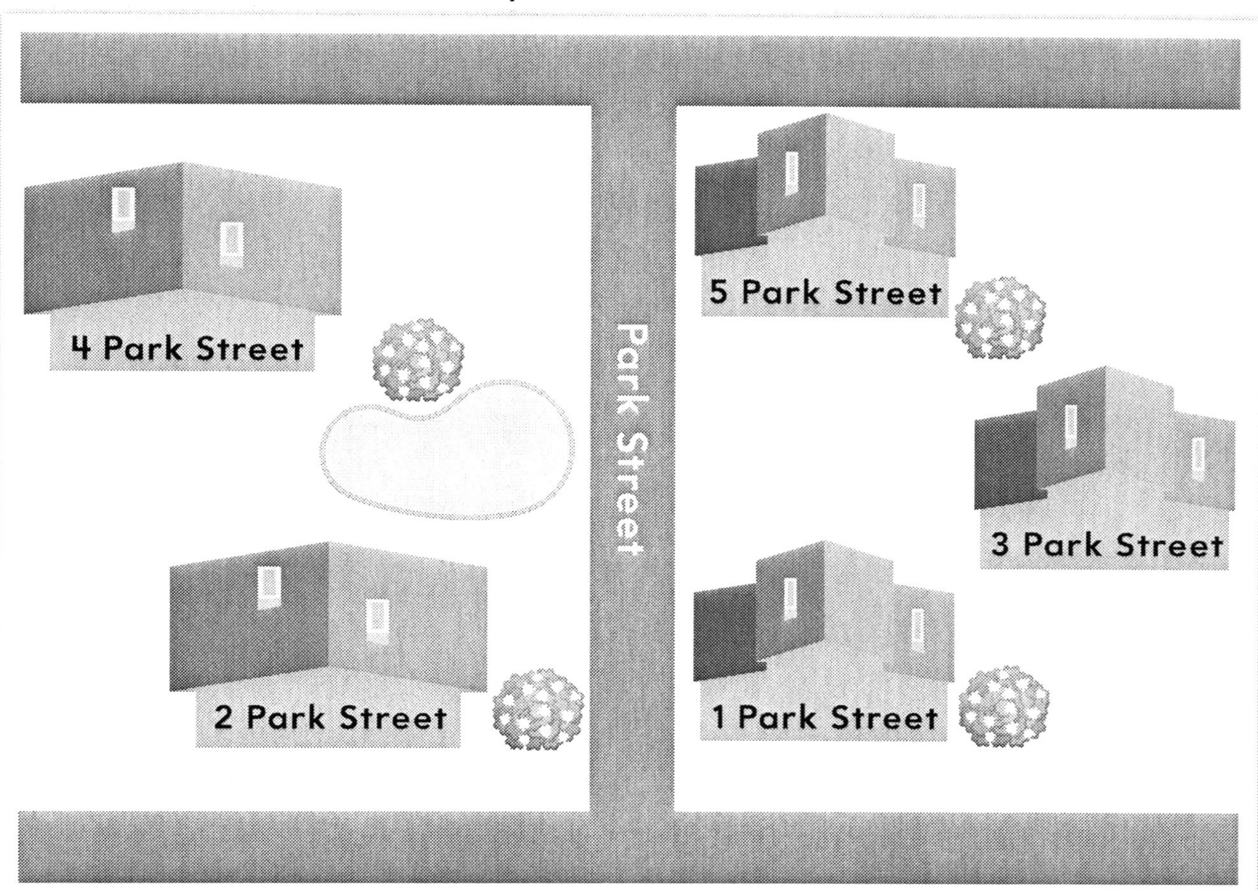

Read About It

Name: ______________________ **Date:** ____________

Directions: Read the text. Study the photo. Then, answer the questions.

Using an Address

Think of a place you want to go. Do you know how to get there? An address would help you get there. It tells people where to find a place. It tells the name of the street. It tells the house number. It also tells the city and state. Homes, stores, and parks all have their own addresses.

1. What information does an address tell?

2. Why does every place have its own address?

Name: ______________________ **Date:** ____________

Directions: Study the addresses on the envelope. Then, answer the questions.

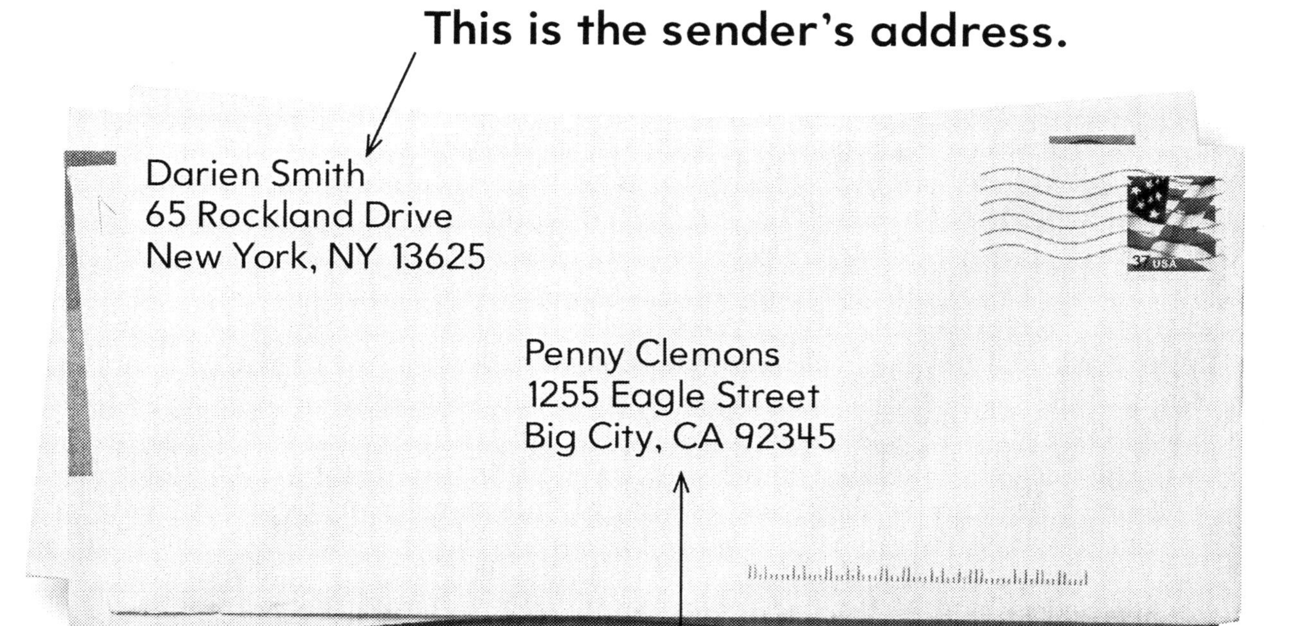

This is the receiver's address. That is the person who will get the letter.

1. What is the name of the person getting the letter?

____

2. Could the letter still be delivered if it were missing the house number? Why or why not?

____

____

Name: ______________________ **Date:** ____________

Directions: Write a letter to a student named Sam Moore. Tell about your school and class. Address the envelope, too. Sam lives at 150 Flower Street.

Red Rock, UT 53245

Name: ______________________ **Date:** ____________

Directions: A route is the way you go to get somewhere. It is the streets you walk on or the path you take. Study the map. Then, answer the questions.t

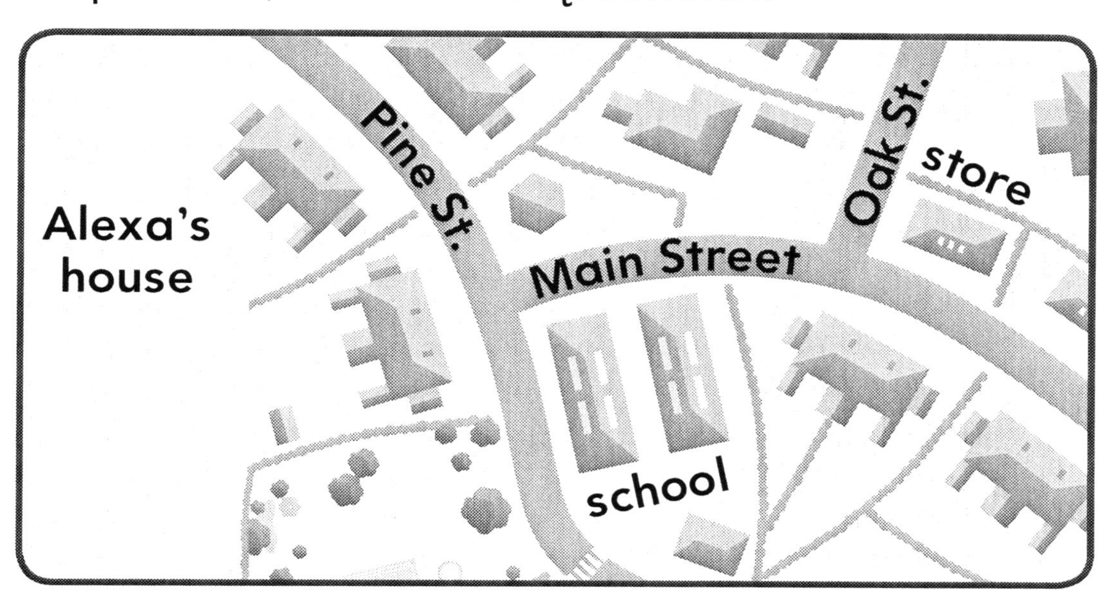

1. Describe the route you would take to get from the school to Alexa's house.

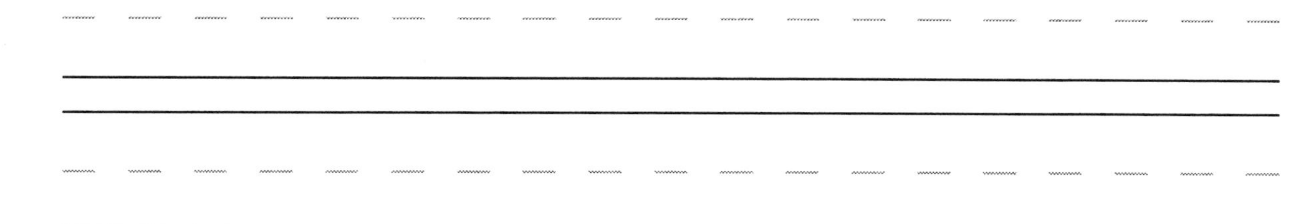

2. Describe the route you would take to get from the school to the store.

Name: ______________________________ **Date:** ______________

Directions: Follow the steps.

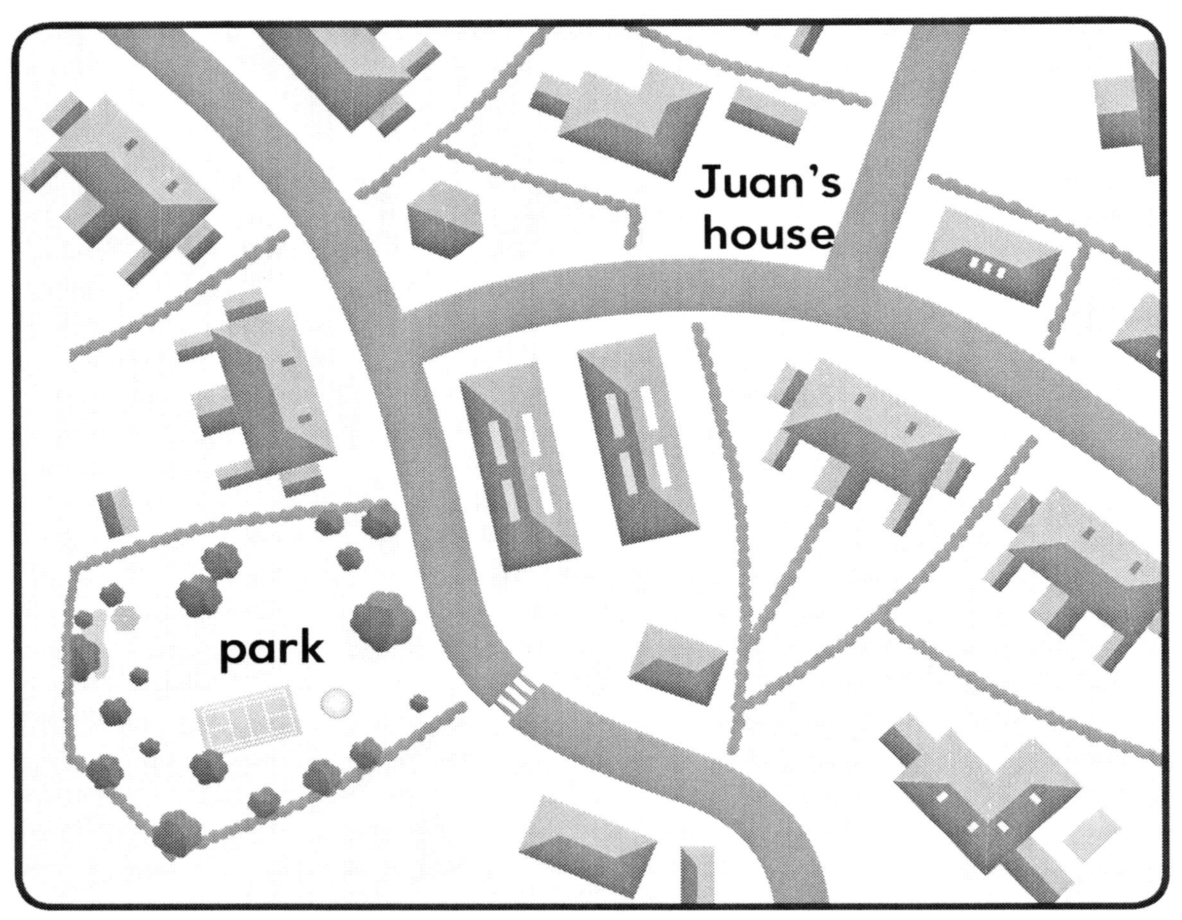

1. Circle Juan's house and the park.

2. Draw a line down the streets to go from Juan's house to the park.

3. Draw an arrow that points to the line.

4. Label the line *to the park*.

Name: ______________________ Date: ____________

Directions: Read the text. Study the photo. Then, answer the questions.

Track It with GPS

Did you know you can map your walk? A digital map on a smart phone will show the route you took. All you need is a GPS device. Smart phones have GPS. And it is used in many ways. Hikers use GPS to track the route they take on trails. Drivers use it to get directions. Some people even put GPS trackers on their pets. Then, if a pet gets lost, its location will show up on the map.

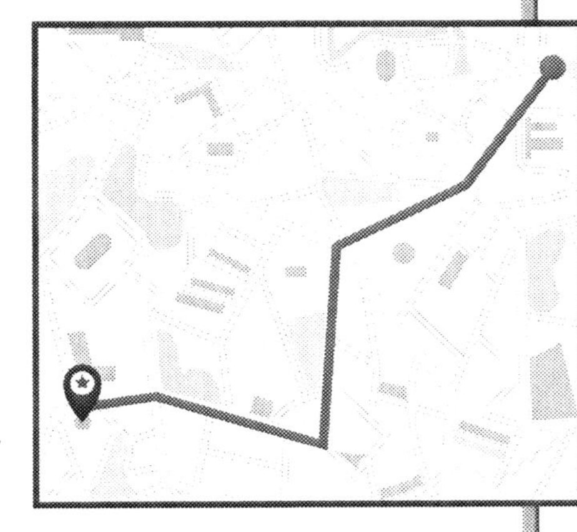

1. What is a route?

____

____

2. What are some ways people use GPS?

____

____

Think About It

Name: ______________________ **Date:** ____________

Directions: A dog is lost in the neighborhood! Study the map. Then, answer the questions.

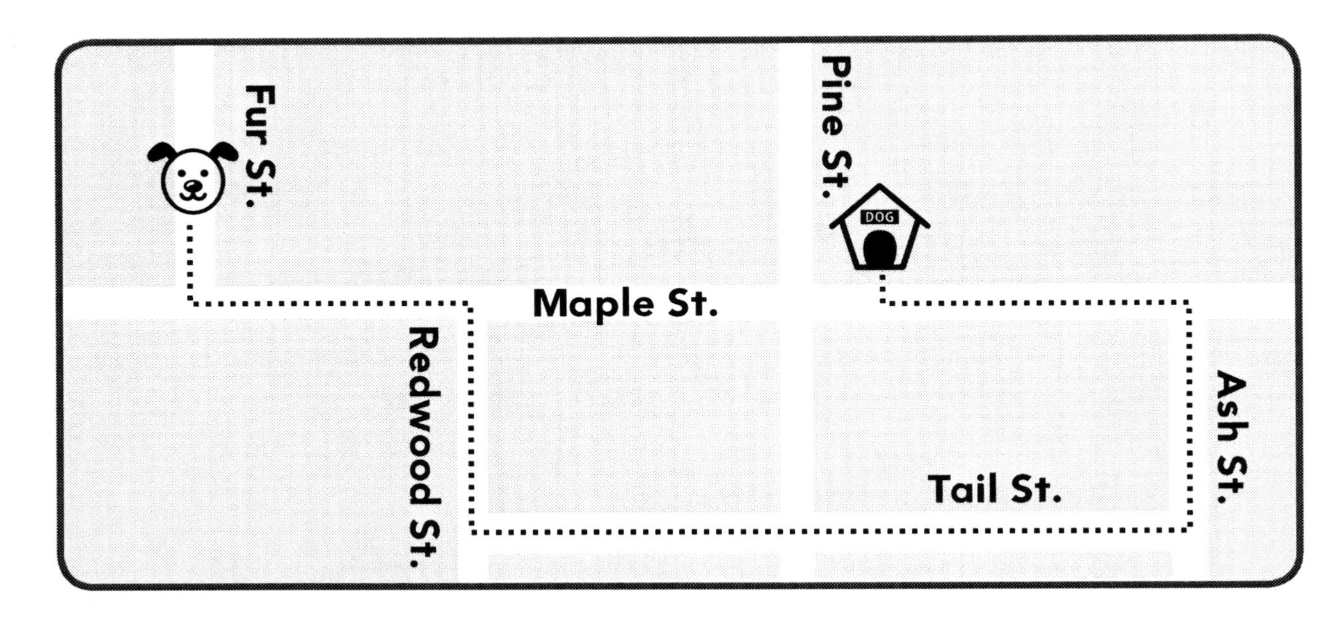

1. Where is the pet located?

____

2. What streets did the dog walk down after he escaped?

____

____

3. Is there a faster route the dog could take to get home? Draw that route onto the map.

Name: ______________________________ **Date:** ______________

Directions: Write a letter to your parents. Tell them how you would use a GPS device. Draw a picture to go with your letter.

____

Dear ____________________,

Name: ______________________ Date: ____________

Directions: Study the globe. Then, answer the questions.

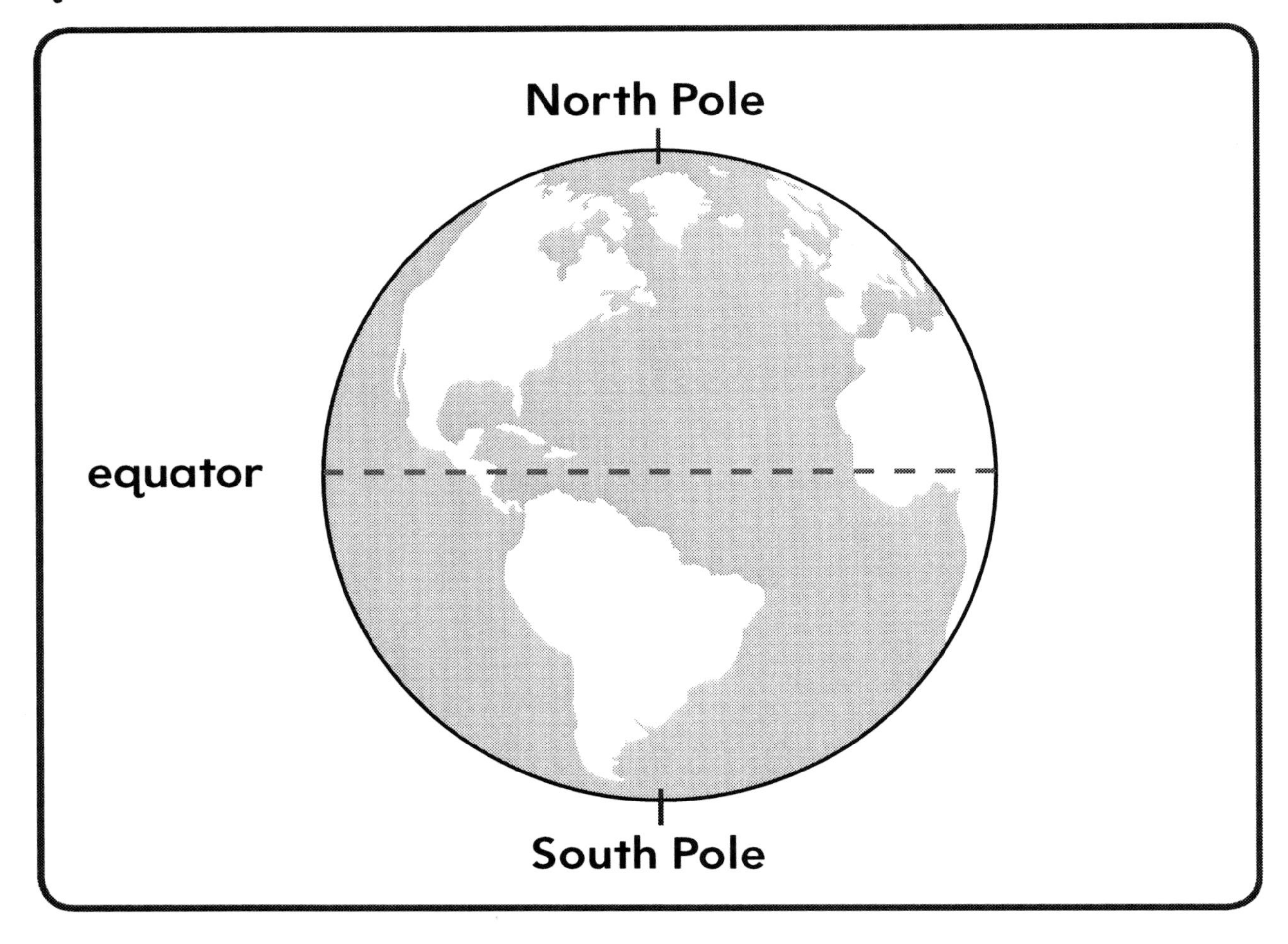

1. Where is the North Pole?

____

2. Where is the South Pole?

____

3. Use a red crayon to trace the equator.

Name: ______________________________ **Date:** ______________

Directions: The equator wraps around the globe. Places close to the equator have warm weather. Follow the steps.

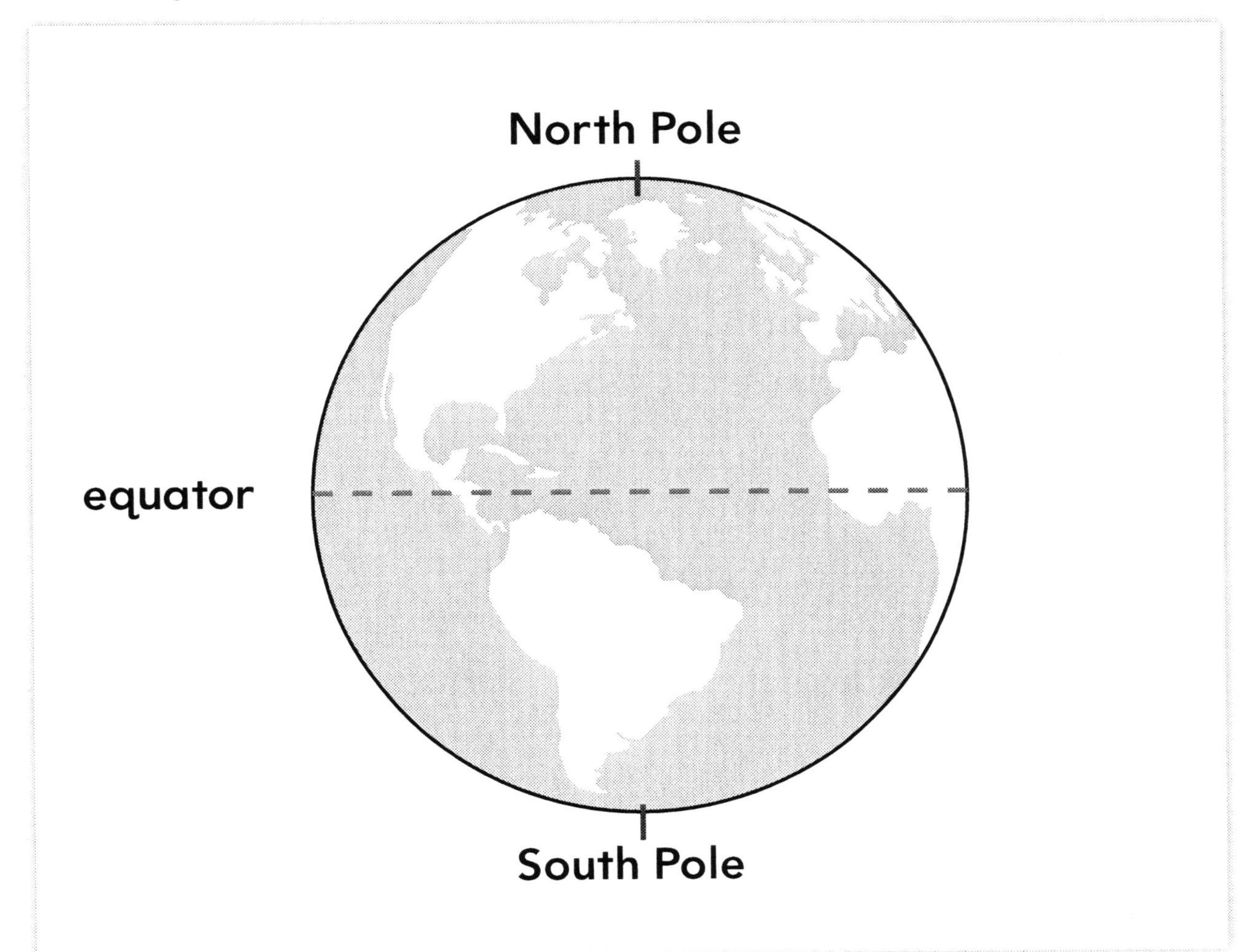

1. Use an orange crayon to shade places you think have warm weather.

2. Use a blue crayon to shade places you think have cooler weather.

Name: ______________________ Date: ____________

Directions: Read the text. Study the photo. Then, answer the questions.

The Poles

It is a hot summer day. But in a different part of the world, it is cold. Why is the weather so different on the same day? It has to do with the sun. The sun shines less directly on the North and South Poles. That means places close to the poles are colder for much of the year. But at the equator, it stays warm. That part of Earth gets more direct sunlight. So most places close to the equator have warm weather for much of the year.

1. What part of Earth gets more sunlight?

____

2. The weather is warmest at the

____

Name: ______________________ Date: ____________

Directions: Draw a line from each picture where it belongs.

North and South Poles **Equator**

1. How did you know where each picture belongs?

____

____

2. Draw other things you might see at the equator.

Name: ______________________________ **Date:** ______________

Directions: Would you rather live closer to the equator or to one of the poles? Why? Draw and write your answer.

Name: ______________________ Date: ____________

Directions: The photo shows two types of communities. One has many homes in it. The other has a lot of open space. Study the photo. Then, answer the questions.

1. Draw a line between the two communities. How did you know where to draw the line?

____

____

2. Describe how the communities look different.

____

____

Name: ______________________ Date: ____________

Directions: A rural community has a lot of open space. It has farms and dirt roads. Study the photo. Then, label the farm and a road.

Name: ______________________ Date: ____________

Directions: Read the text. Study the photo. Then, answer the questions.

Wheat on the Plains

Farmers live and work in rural communities. Here is a farmer at work on his wheat farm. He is driving a machine that cuts and processes the wheat. Wheat grows on the plains. The plains are flat areas of land. We use the wheat to make food. Wheat can be used to make bread and baked goods. It can be used to make cereal, too. Pasta also comes from wheat. Do you eat any of those foods?

1. Describe the land in the photo.

____

____

2. What can wheat be used to make?

____

____

Think About It

Name: ______________________ **Date:** ____________

Directions: Where does each picture belong? Draw a line from each picture to the correct box on the left.

not rural	
rural	

1. How do you know which things belong in a rural community?

2. Add pictures of other things in a rural community.

Name: ______________________________ **Date:** ____________

Directions: Draw your favorite food made from wheat. Then, list the other food you eat that comes from wheat.

Name: ______________________ **Date:** ____________

Directions: Study the map. Then, answer the questions.

1. Describe what you see on this map.

____

____

2. Circle a forest.

3. Draw symbols that stand for a forest, a river, a mountain, and a lake.

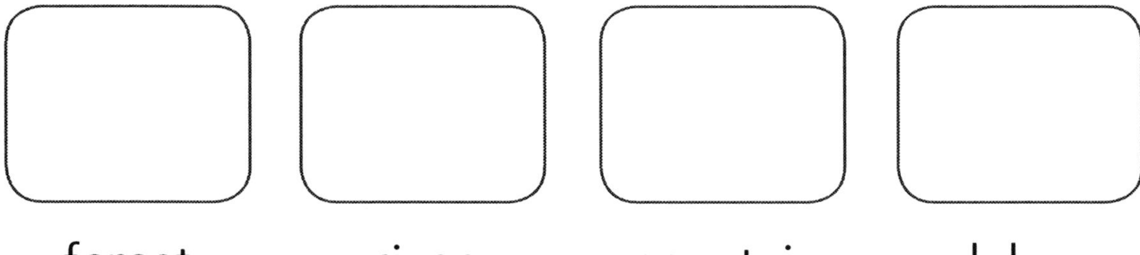

forest river mountain lake

Name: ______________________________ Date: ______________

Directions: Complete the map key. Write what each symbol stands for.

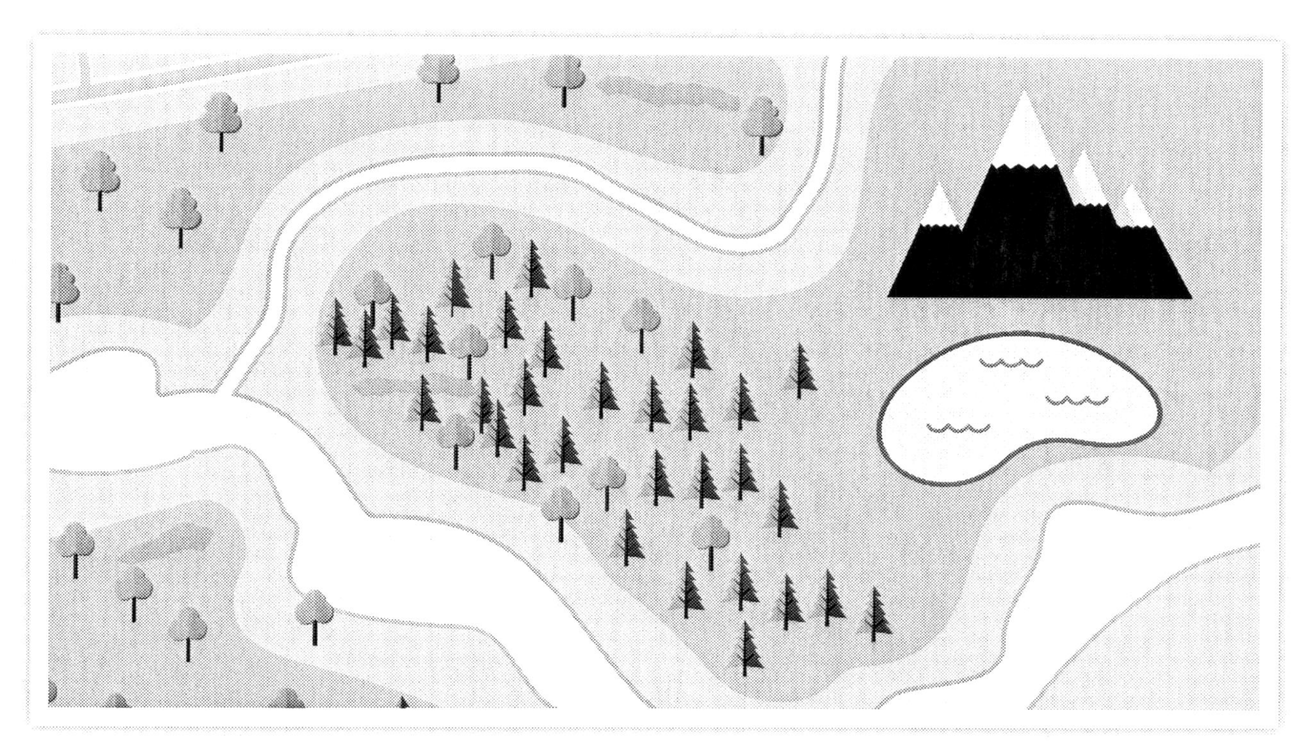

Key	
1.	
2.	
3.	
4.	

Read About It

Name: ______________________ **Date:** ____________

Directions: Read the text. Study the photo. Then, answer the questions.

Landforms

A landform is made by nature. There are many types. A mountain is land that rises up and is tall. Hills also rise up, but they are not as tall. In some places, the land is flat and does not rise up. That flat land is called *the plains*.

Plateaus are a different kind of flat land. They rise up like mountains. But they are flat only on the top. What landforms have you seen?

1. What is the difference between the plains and a plateau?

____

____

2. How is a plateau like a mountain?

____

____

Name: ______________________ **Date:** ____________

Directions: This picture shows landforms. Study the picture. Then, answer the questions.

1. Choose one landform from the picture. What is it called? Tell what it looks like.

____

____

2. Start at the mountain. Then, trace your finger along the river. Tell a friend what it leads to.

3. How are mountains and hills different?

____

____

Name: ________________________________ **Date:** ______________

Directions: Think of a landform you have seen in real life. Complete the chart to tell about it.

Name of Landform	Where You Saw It
Describe it.	
Draw it.	

Name: ______________________________ Date: ______________

Directions: Study the map. Then, answer the questions.

Resources

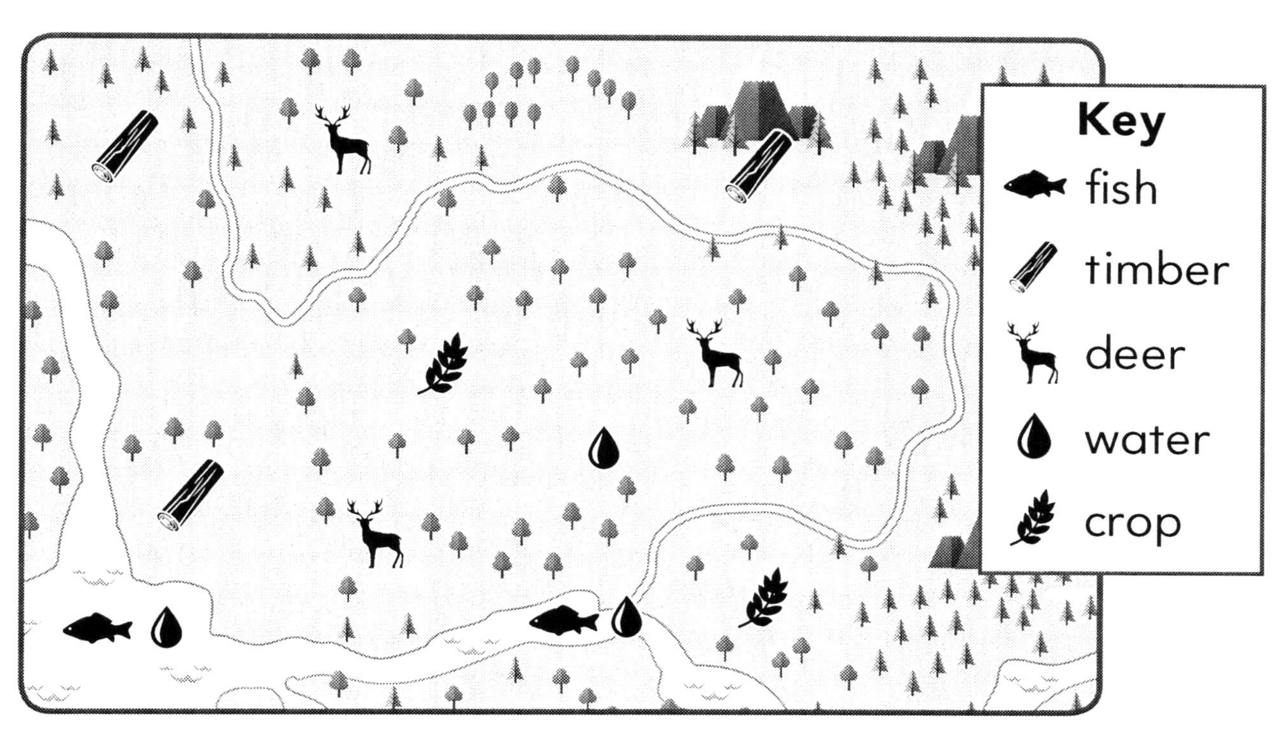

1. What resources are shown on the map?

____

____

2. What resources can people get from the river?

____

____

Name: ______________________________ Date: ____________

Directions: Make a map key. It should show the resources found in this area.

Resources

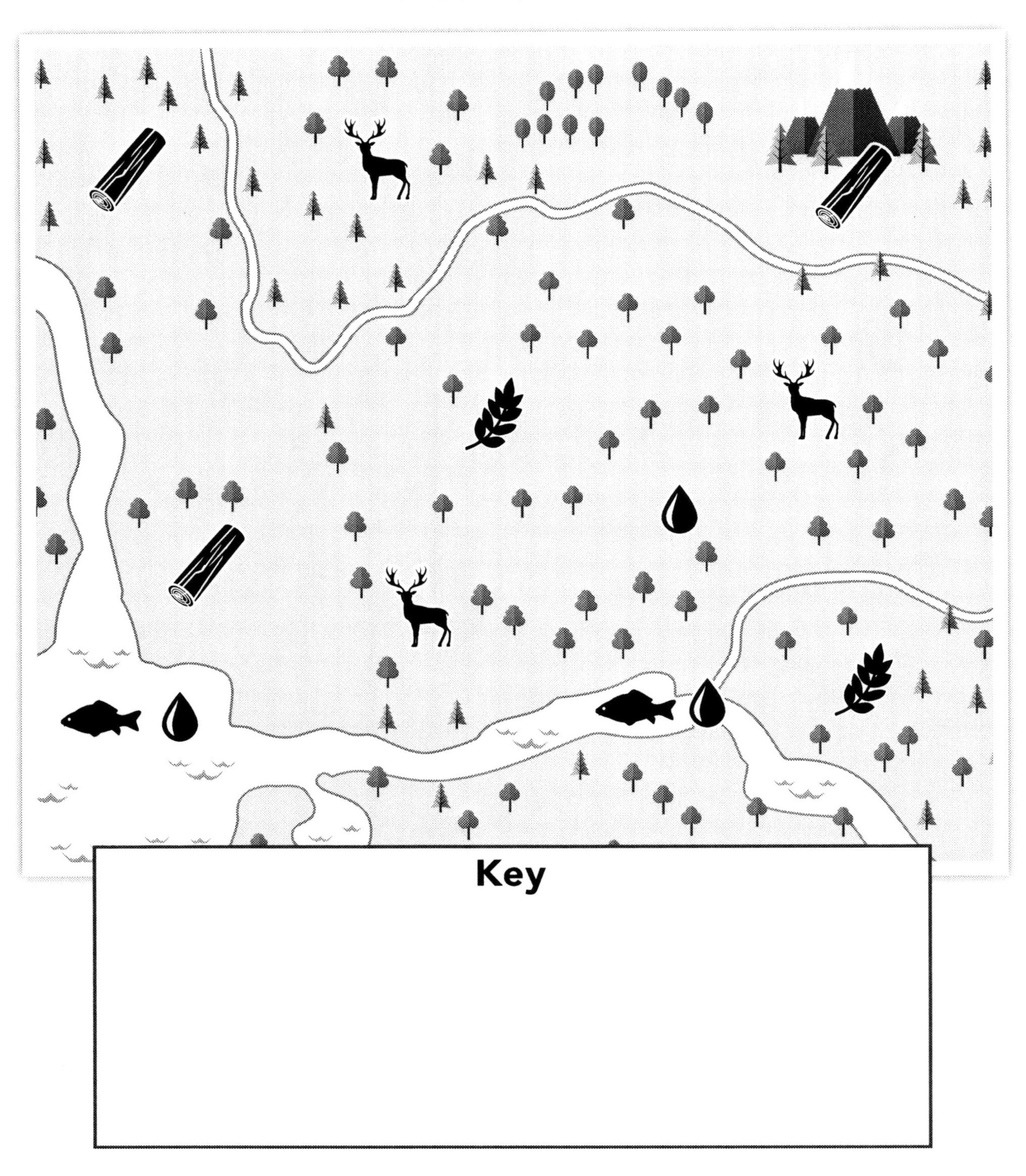

Name: ______________________ Date: ____________

Directions: Read the text. Study the photo. Then, answer the questions.

Timber

Timber is a renewable resource. That means it will not run out. We can always plant more trees for timber. First, the trees are grown. Then, they are chopped down. A truck takes them to a sawmill. Then, the wood products are bundled up. They are sent to factories and stores. They are turned into furniture and tools.

1. What happens right after a tree is chopped down?

2. Explain why timber is a renewable resource.

Name: ______________________________ **Date:** ______________

Directions: Draw a line from each picture to the correct box. Then, answer the questions.

table

toaster

wood products

pan

oven mitts

pencil

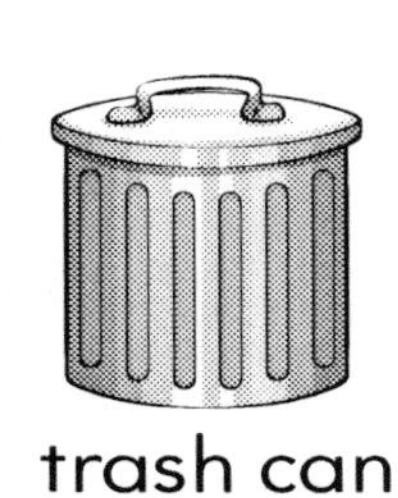

trash can

not wood products

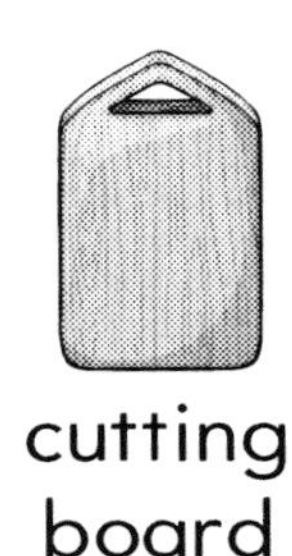

cutting board

spoon

1. Which of these wood products do you have at home?

____

____

2. Name a wood product that can be made from something else.

____

Name: ______________________ **Date:** ____________

Directions: Draw and write about something in your home that is made from wood.

Name: ________________________ **Date:** ______________

Directions: Study the map. Then, answer the questions.

Bodies of Water

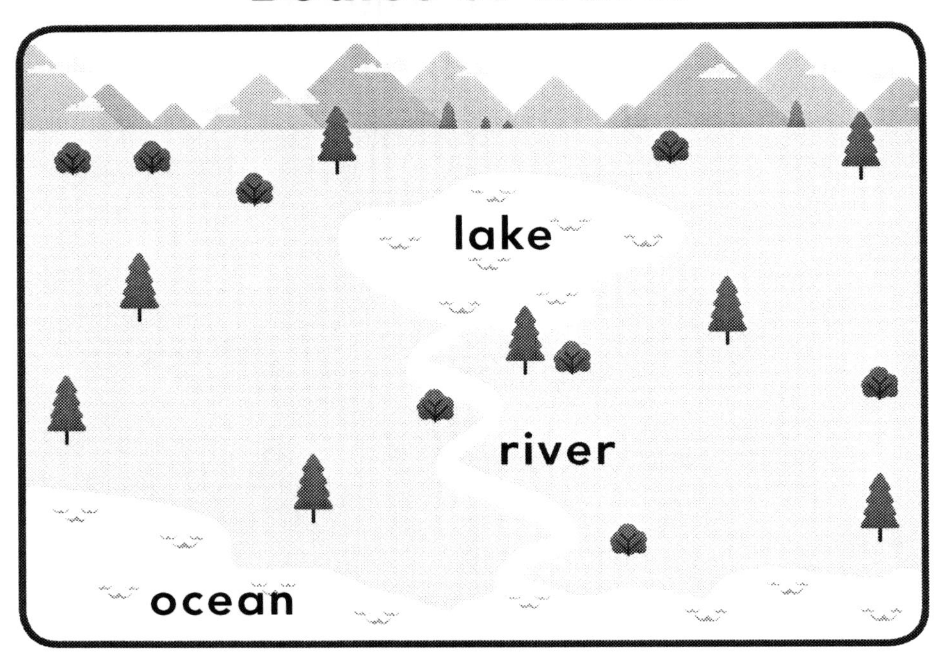

1. Describe the size of the ocean compared to the size of the lake.

____

____

2. How is the shape of a river the same or different from the shape of a lake?

____

____

Name: ______________________________ **Date:** ______________

Directions: Label the bodies of water. There is a lake, an ocean, and a river. Then, color the bodies of water blue.

Bodies of Water

Name: ______________________________ **Date:** ______________

Directions: Read the text. Study the photo. Then, answer the questions.

The Deepest Lake

Lake Baikal is the deepest lake in the world. In one part of the lake, it is more than 5,250 feet (1,600 m) deep. It holds more water than any other lake on Earth. It is also the oldest lake in the world. Lakes come in all kinds of shapes. This one is shaped like a long crescent. Water flows into it from rivers. All lakes are surrounded by land. Not all lakes have islands in the middle. Lake Baikal has 27!

1. How does water get into a lake?

____

____

2. What can be found in the middle of Lake Baikal?

____

____

Think About It

Name: ______________________________ **Date:** ______________

Directions: This is a map of Lake Baikal. Study the map. Then, answer the questions.

1. Trace a river that carries water into Lake Baikal.

2. How do you know land surrounds the lake?

____

____

Name: ______________________________ **Date:** ______________

Directions: Would you rather sail across an ocean, down a river, or around a lake? Why? Draw and write your answer.

Name: ______________________ **Date:** ____________

Directions: This is a photo of the Grand Canyon in Arizona. Study the photo. Then, answer the questions.

1. Describe what a canyon looks like.

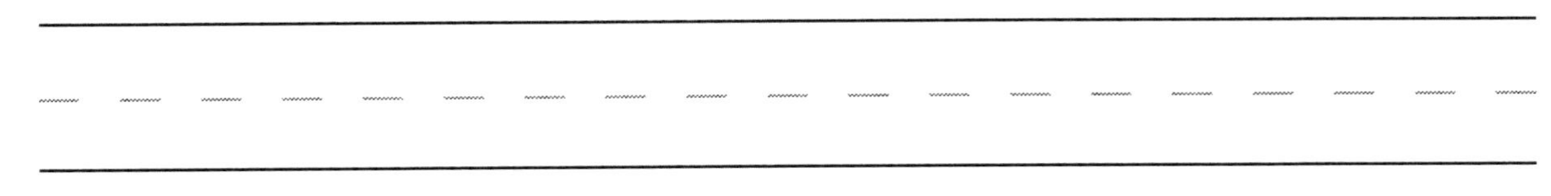

2. What is at the bottom of the canyon?

Name: ______________________ Date: ____________

Directions: Study the photo. Follow the steps.

1. Draw an arrow pointing to the river.
2. Label the canyon walls.
3. Label the rim of the canyon. The rim is the top of the canyon.
4. Tell a friend about the Grand Canyon.

Name: ______________________ **Date:** ______________

Directions: Read the text. Study the photo. Then, answer the questions.

China's Grand Canyon

There is a Grand Canyon in China. It is three times deeper than the one in the United States! People drive or fly from far away just to see it. A river is powerful. It can cut through rocks. That is how this Grand Canyon formed. The river cut a path through the land. It made a deep canyon. That canyon got deeper as time went on. And now, it is the deepest in the world.

1. Why does it say the river is powerful?

____

____

2. How does a canyon form?

____

____

Think About It

Name: ______________________ Date: ____________

Directions: The photos show China's Grand Canyon and the one in the United States. Study the photos. Then, answer the questions.

United States

China

1. Describe the walls of both canyons.

____

____

2. Describe how the canyons are alike and how they are different.

____

____

____

Name: ______________________ **Date:** ____________

Directions: Would you rather visit the Grand Canyon in the United States or China? Why? Draw and write your answer.

Name: ______________________ **Date:** ____________

Directions: This map shows where wind is used for power. Study the map. Answer the questions.

Wind Power

1. Draw the wind turbines symbol.

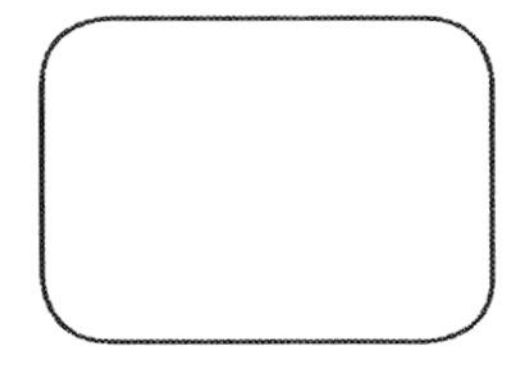

2. What other symbols appear on the map?

____

____

3. Tell a friend where the wind turbines are on the map.

Name: ______________________________ Date: ______________

Directions: A new wind farm is being built. Follow the steps.

Wind Power

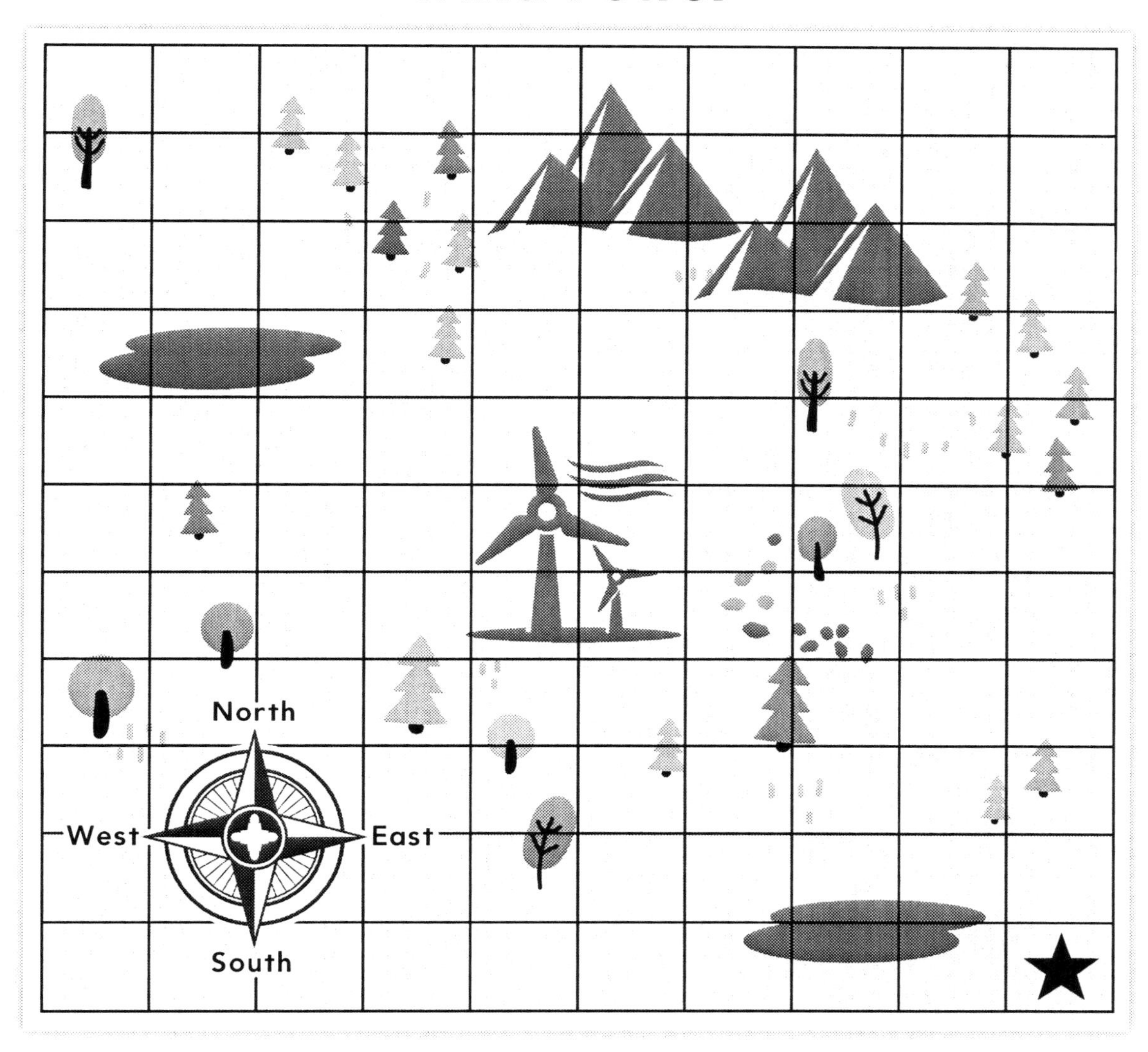

1. Start at the star.
2. Move eight squares west.
3. Move nine squares north.
4. Draw a wind turbine here.
5. Tell a friend what the new wind farm is near.

Read About It

Name: ______________________ **Date:** ____________

Directions: Read the text. Study the photo. Then, answer the questions.

Wind Farms

This is a wind farm. Those tall structures are wind turbines. The blades spin when there is a lot of wind. When the blades spin, they capture the wind's energy. People can use that energy in their homes. Wind is a renewable resource. That means it will never run out. Wind farms are built in the places with strong and steady wind.

1. Where are wind farms built?

____

____

2. Why is wind a renewable resource?

____

____

Name: ______________________ Date: ____________

Directions: A city votes to build a wind farm. They cannot agree where to put it. Study the table. Then, answer the questions.

Places	Wind Speeds
meadow	22 miles per hour
high hill	3 miles per hour
low hill	14 miles per hour

1. Which place has the highest wind speeds?

____

2. Explain why the wind farm should not be built on the high hill.

____

____

____

Name: ______________________________ Date: ______________

Directions: How would you feel about a wind farm being built in your community? List the positives and negatives on the chart.

Positives	Negatives

Name: ______________________ **Date:** ______________

Directions: This is a map of an urban community. Study the map. Then, answer the questions.

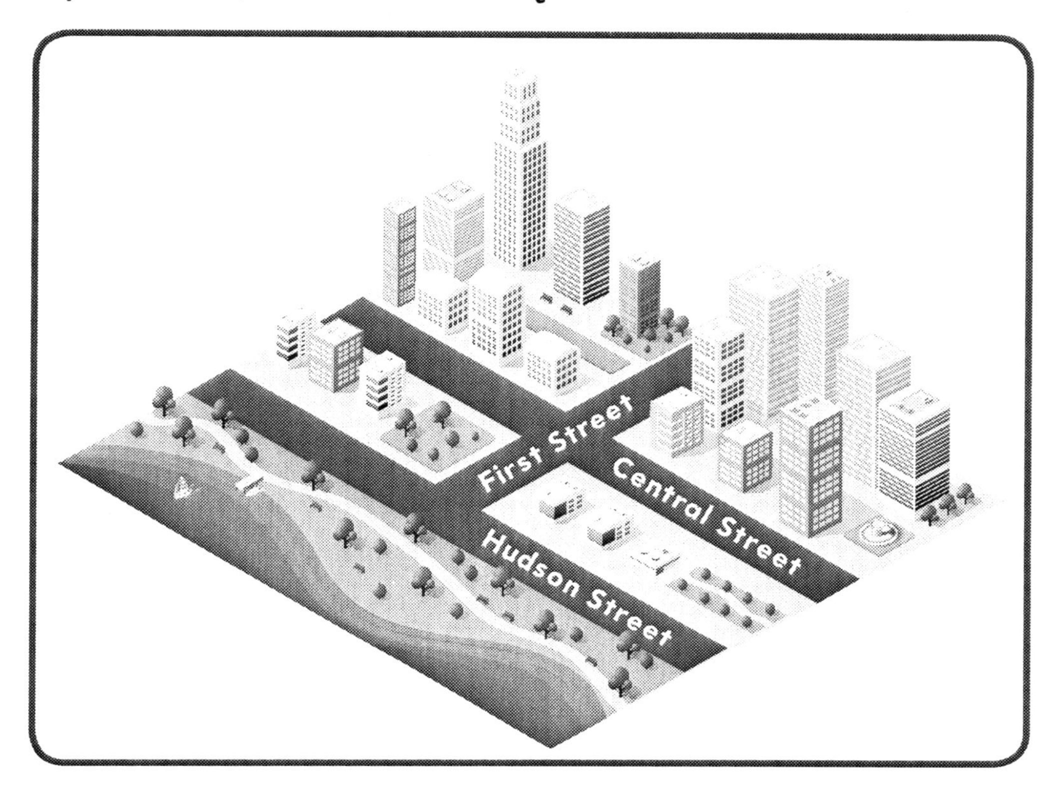

1. Draw a star in the park. Draw a tall antenna on top of a skyscraper.

2. What street could you drive down to pass the lake?

____

3. What street starts at the park and goes through the city?

____

Name: ______________________ **Date:** ____________

Directions: A city has tall buildings. It has parks, stores, streets, cars, and a lot of people. Draw a city in the box.

Try It! Tell a friend about the things you drew on the map.

Name: ______________________________ **Date:** ____________

Directions: Read the text. Study the photo. Then, answer the questions.

City of Smog

Shanghai, China, is one of the world's largest cities. It is an urban community. It is crowded. People work in the skyscrapers. They live in tall buildings, too. There is a lot to do there.

This city has a lot of smog. That can make it hard to see and breathe outside. It can be so bad that some people need to wear masks. That keeps them from breathing in the smog.

1. What makes Shanghai an urban community?

____

____

2. How does smog affect the people who live there?

____

____

Name: ______________________ Date: ____________

Directions: This photo shows some of the ways people get around in a city. Study the photo. Then, answer the questions.

1. List the ways people travel around the city.

2. Explain why taking a bus helps improve traffic.

Name: ______________________ Date: ____________

Directions: Draw and write about a place you like in your community.

Reading Maps

Name: ______________________ Date: ____________

Directions: Study the map. Then, answer the questions.

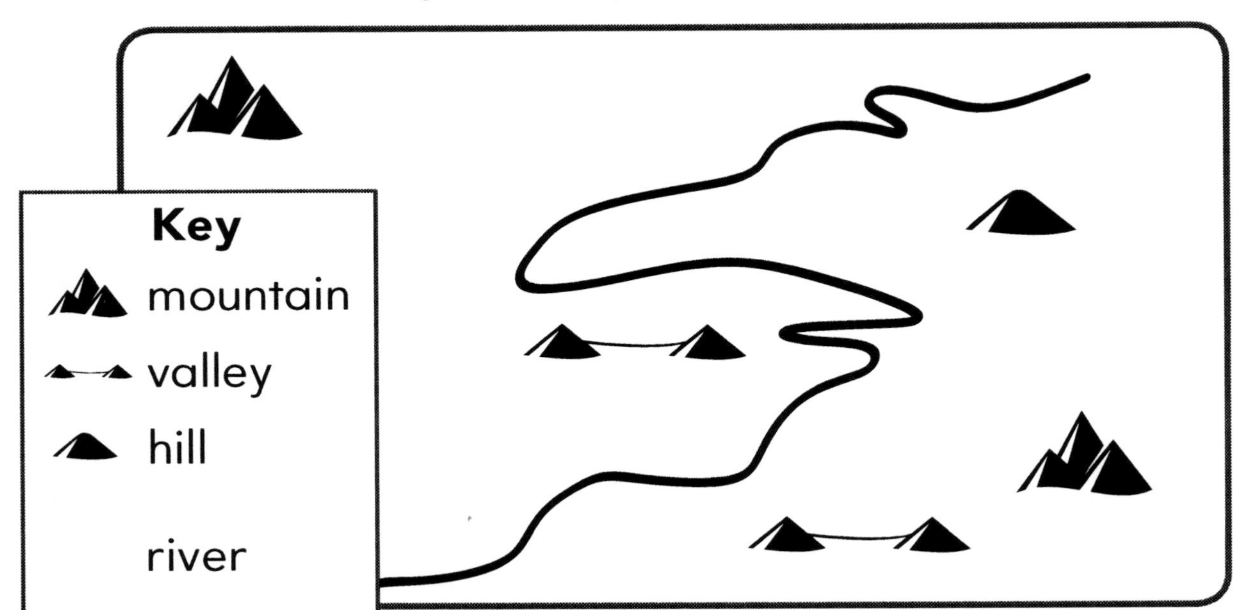

1. Draw the symbol that stands for a river in the key.

2. Circle the symbol that stands for a mountain.

3. Study the mountain and hill symbols. How are they different?

Name: ______________________________ **Date:** ______________

Directions: Follow the steps.

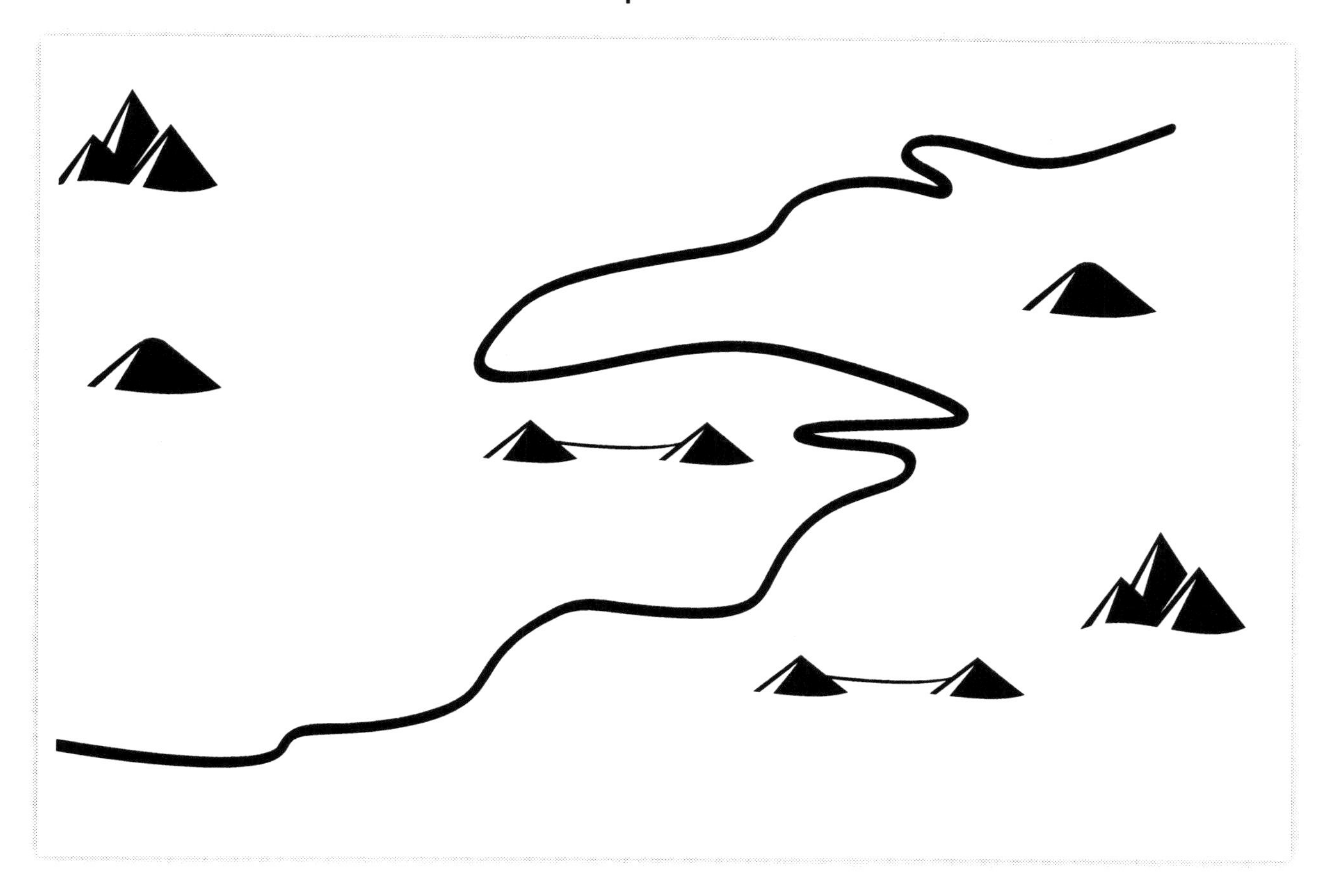

1. Trace the path of the river in blue.

2. Draw a symbol to stand for a gold nugget.

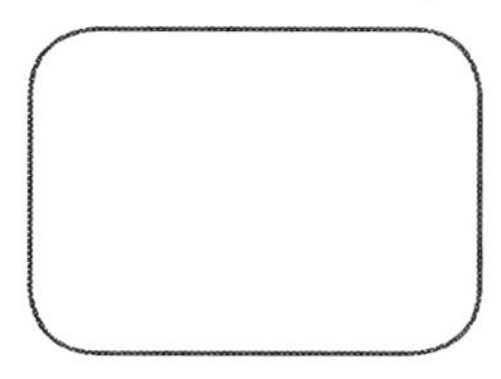

3. Add your nugget symbol to the map in a few places. Draw it near the mountains and the river.

4. Tell a friend about your symbol.

Name: ______________________________ **Date:** ______________

Directions: Read the text. Study the photo. Then, answer the questions.

A Non-Renewable Resource

People get resources from Earth. Some of those resources will never run out. Some of them can run out. Gold is a non-renewable resource. That means it can run out. Gold is a shiny, yellow metal. It is rare. That means it is hard to find. That makes it worth a lot of money. Gold is found in solid rock. It is also found in broken up pieces of rock in a stream.

1. Why is gold a non-renewable resource?

____

____

2. Where can gold be found?

____

____

____

Name: ______________________ **Date:** ____________

Directions: The graph shows different ways gold is used. Study the graph. Then, answer the questions.

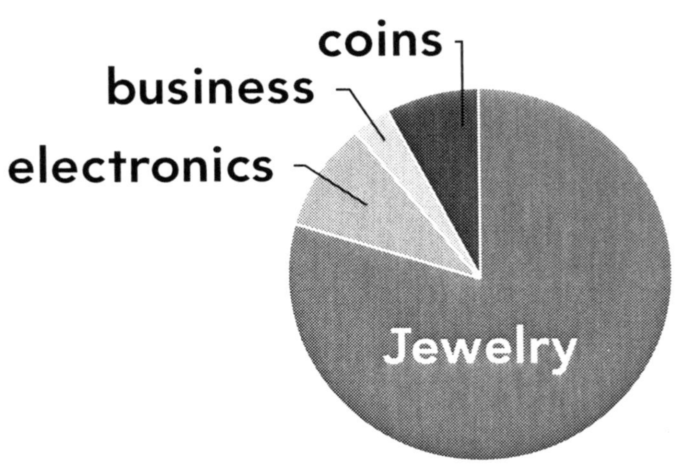

1. People use gold the most to make

2. What are other ways people use gold?

3. Study the many uses of gold. Which one do you think is the most important. Why?

Name: ______________________ **Date:** ____________

Directions: Draw and color a piece of gold jewelry. Then, write about who you would give this to and why.

Name: ______________________________ **Date:** ______________

Directions: This city is made up of islands. The white lines are water. They are called *canals*. Study the map. Then, answer the questions.

Venice

1. How can you tell the land is made of islands?

____

____

2. What are some ways people could get from one island to the other?

____

____

Creating Maps

Name: ______________________________ **Date:** ______________

Directions: Follow the steps.

Venice

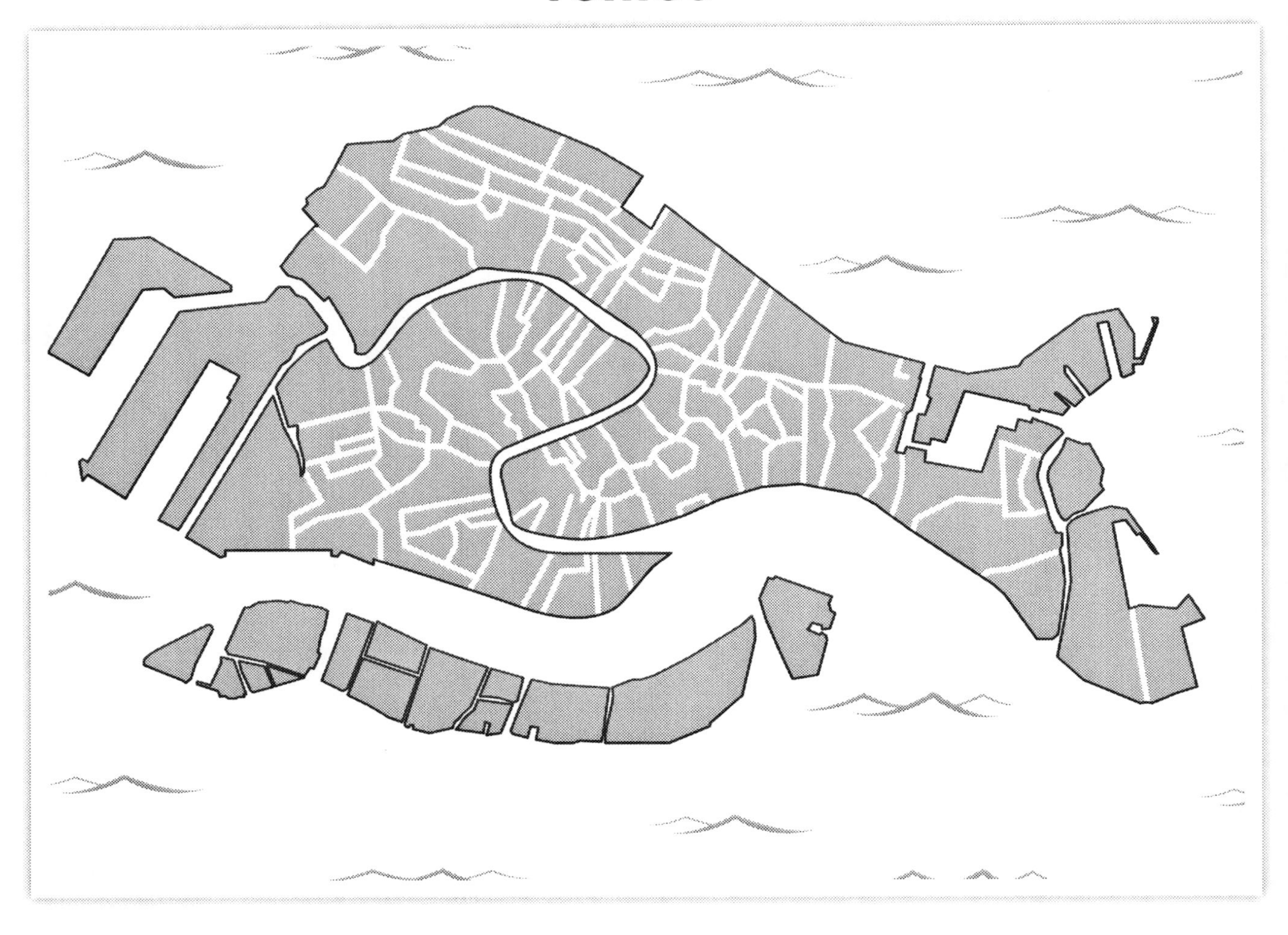

1. The Grand Canal runs through the middle of Venice, Italy. It is shaped like a backward *S*. Shade it blue on the map.
2. Label the Grand Canal.
3. Trace five other canals in blue.
4. Color the water around the islands blue.

Name: ________________________ **Date:** ____________

Directions: Read the text. Study the photo. Then, answer the questions.

Venice

Venice is a city in Italy. There are 118 islands in Venice. Most cities have streets to get from place to place. Venice has canals. The canals hold water. They are wide enough for small boats. The Grand Canal is the widest one. People can catch a water bus there, too. People use bridges to get from one island to the next. They can cross the bridge by foot or bike.

1. How can people get around Venice?

____

____

2. Explain why the city needs bridges.

____

____

Name: ____________________ Date: ____________

Directions: Draw a route from the circle to the star. Use the canals and bridges.

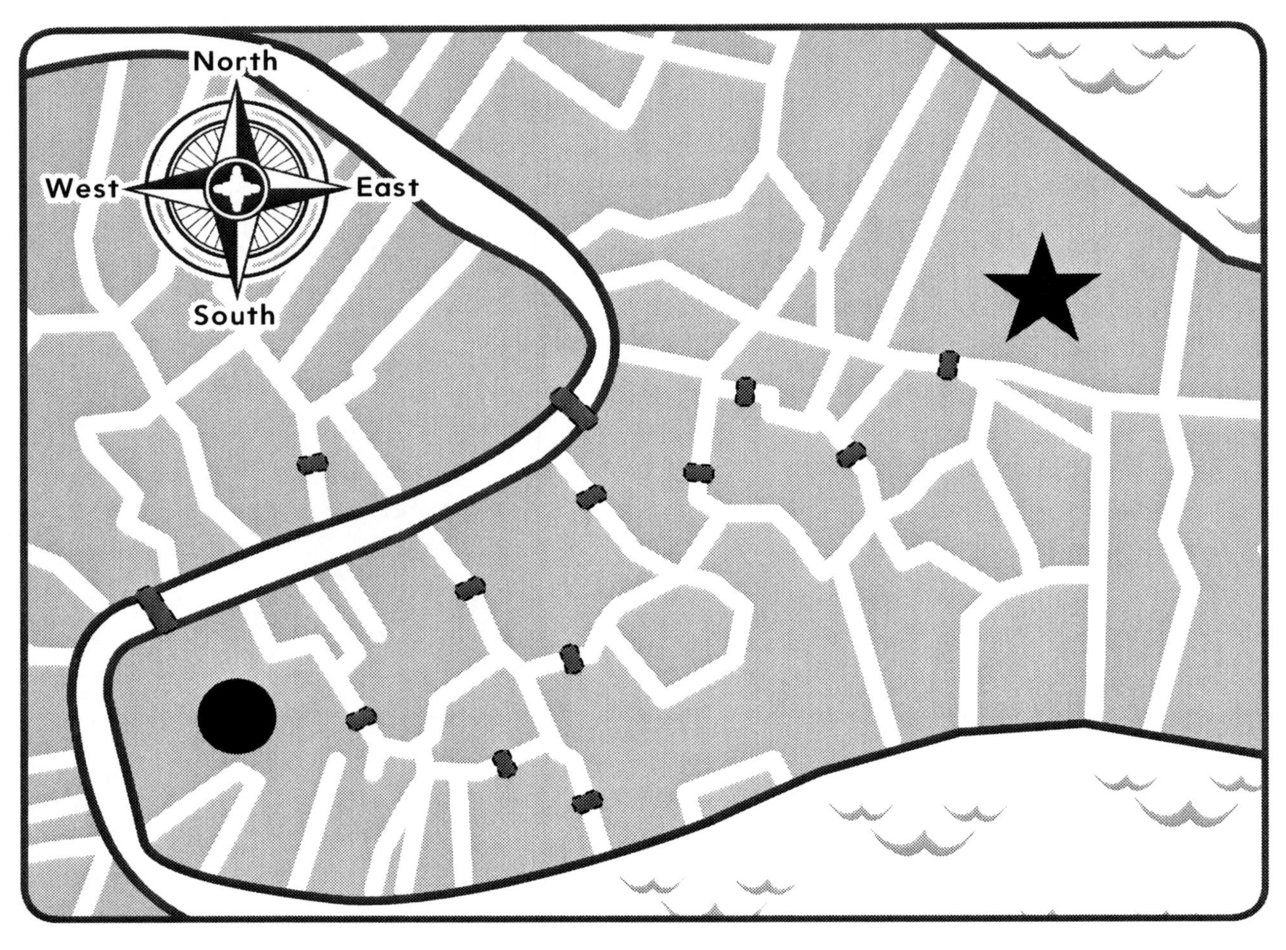

1. How many canals are in your route?

____

2. Use cardinal directions to describe your route.

____

____

Name: ______________________ **Date:** ____________

Directions: Draw and write about how you would travel around Venice.

Reading Maps

Name: ______________________________ **Date:** ______________

Directions: A landmark is something that stands out in a place. It is often big. People can see it from far away. Study the map. Then, answer the questions.

1. Describe a landmark on the left side of the map.

____

____

2. Circle the largest building. Could it be seen from far away? Tell a friend how you know.

Name: ______________________________ **Date:** ______________

Directions: Follow the steps.

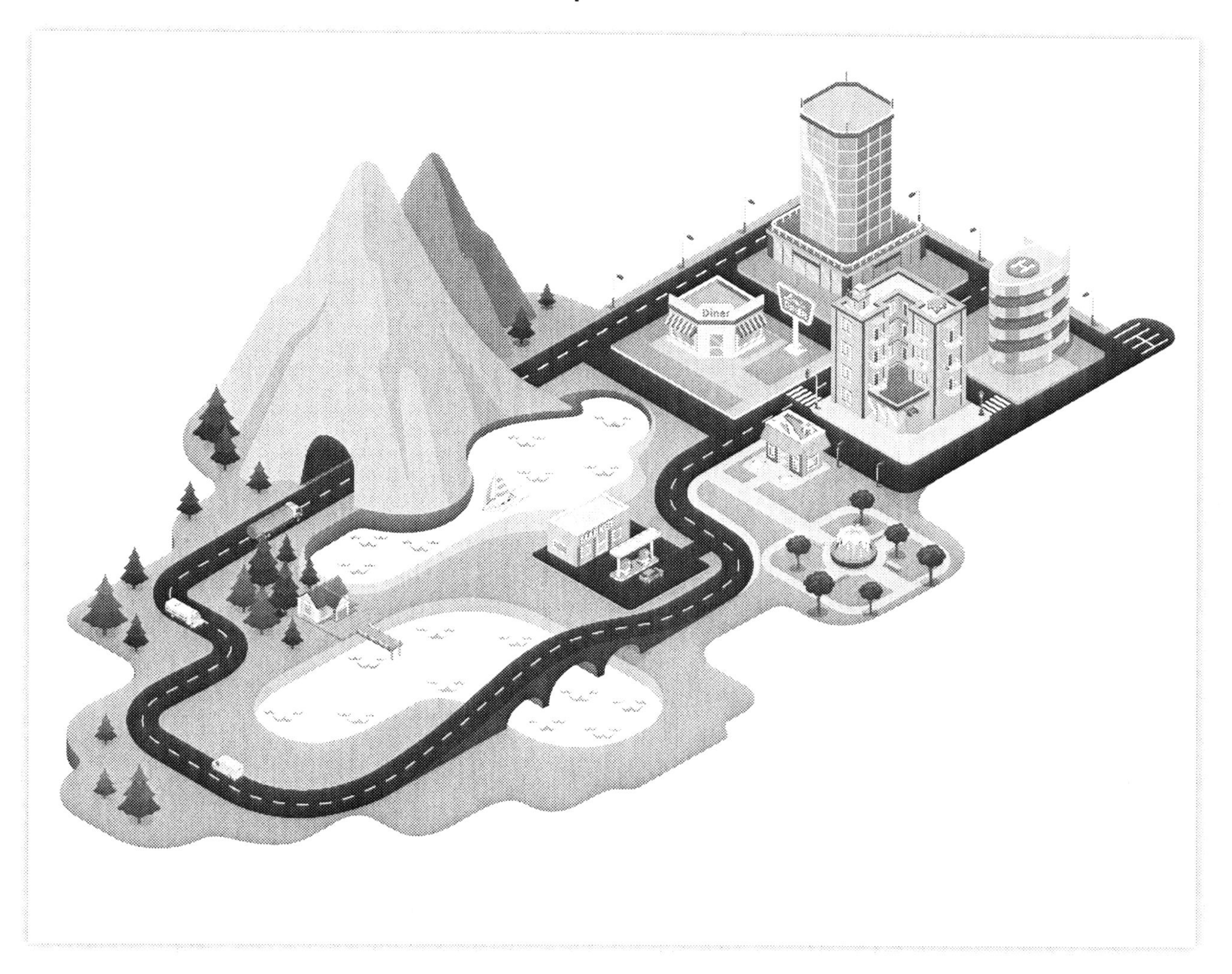

1. Draw something on the map that could be seen from far away.

2. Tell a friend why this would be a landmark.

3. Circle all the other landmarks on the map.

4. Tell a friend which landmark you think is the most unique.

Read About It

Name: ______________________ **Date:** ____________

Directions: Read the text. Study the photo. Then, answer the questions.

Landmarks

Landmarks can be made by nature, such as waterfalls or hills. Or they can be made by people, such as special buildings or statues. People can see a landmark from a distance. It stands out. It helps us know where we are. Big Ben is a landmark in London. A friend might say, "Let's meet at the café near Big Ben." Right away, the other friend will know where to meet.

1. What is a landmark?

____

____

2. Explain why Big Ben is a landmark.

____

____

Name: ______________________ Date: ____________

Directions: Study the photo. Then, answer the questions.

Think About It

1. Describe a landmark at this fair.

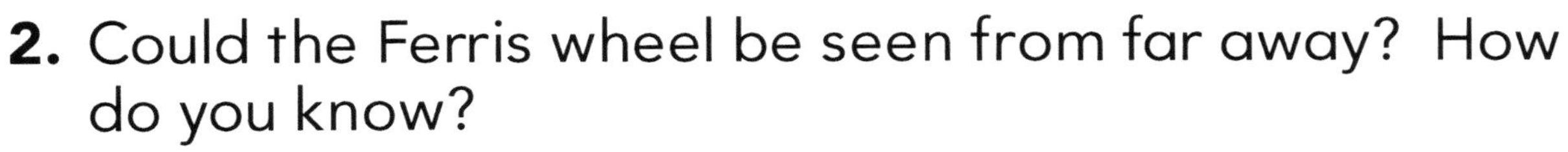

2. Could the Ferris wheel be seen from far away? How do you know?

Name: ________________________ Date: ____________

Directions: Draw and write about a landmark in your community.

Name: ______________________________ **Date:** ______________

Directions: Oil is a black, sticky liquid. When it spills, it makes a huge mess. Study the map. Then, answer the questions.

Oil Spill

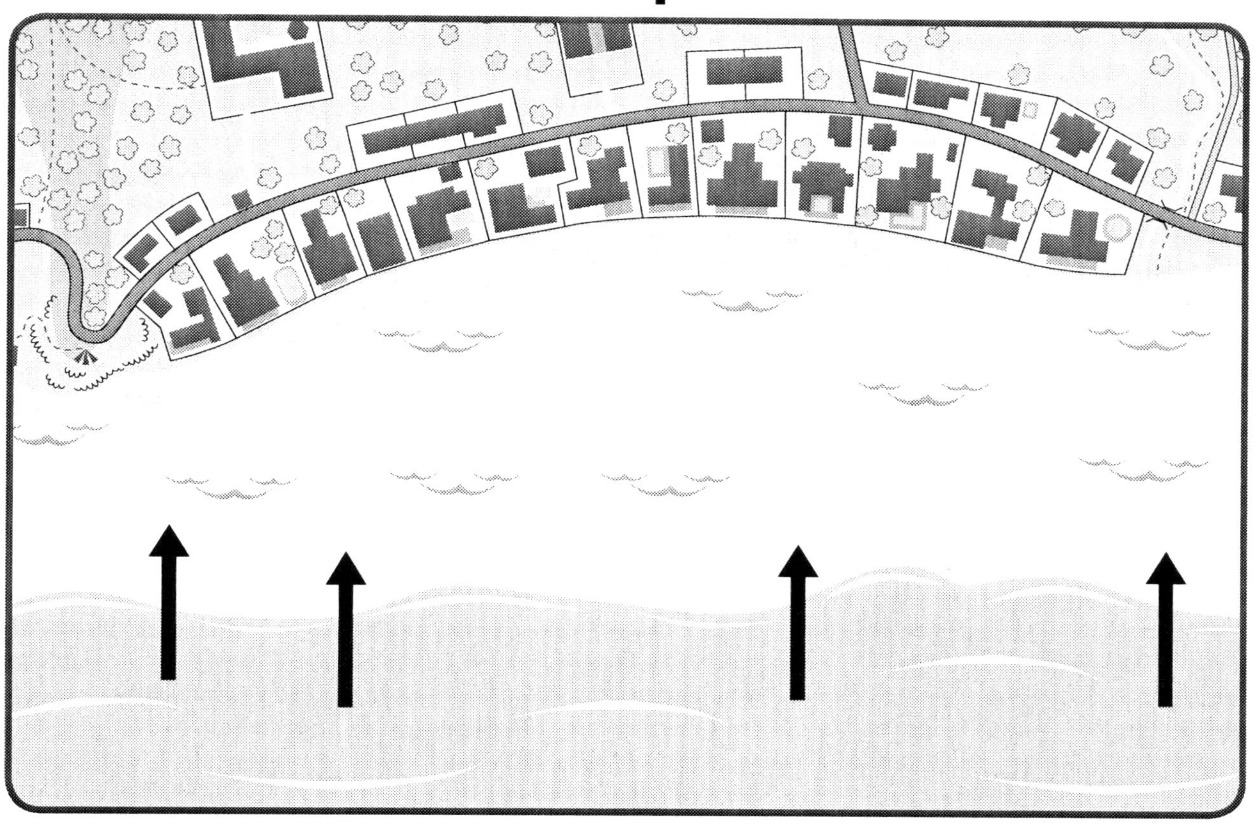

1. Color the ocean blue and the oil spill black.

2. Do you think the oil will reach the shore? Why or why not?

____

____

____

Name: ______________________________ Date: ____________

Directions: The oil spill is moving closer to the beach. How much of the beach will be covered in oil? Follow the steps to find out.

Oil Spill

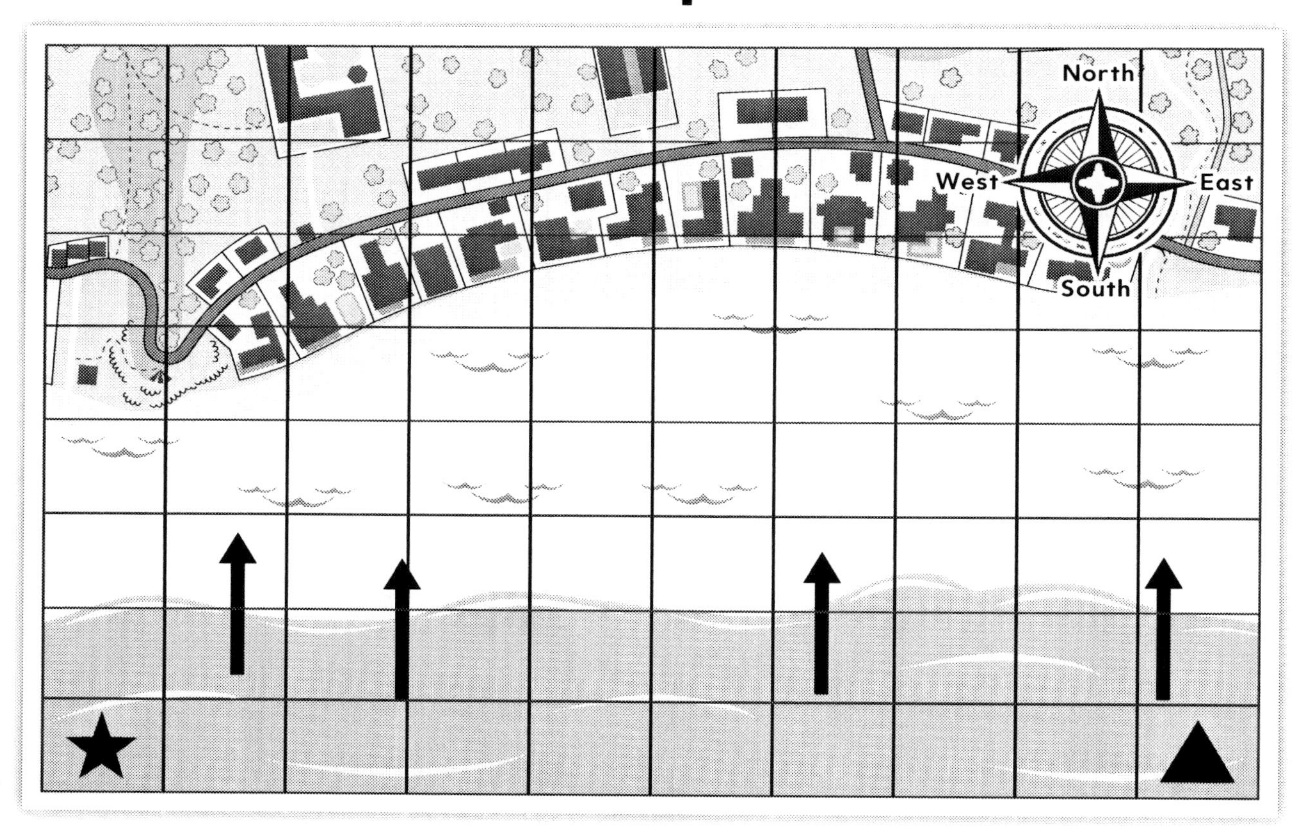

1. Start at the star. Go five squares north and eight squares east.

2. Draw an oil symbol in that spot.

3. Start at the triangle. Go four squares north and eight squares west.

4. Draw an oil symbol in that spot.

5. Use a black crayon to color the beach between the two symbols.

Name: ______________________ Date: ____________

Directions: Read the text. Then, answer the questions.

Oil

Oil is a sticky, gooey liquid. We use it to make fuel. Cars and planes run on fuel. Oil is a non-renewable resource. There is only so much of it in the world. Oil is pumped out of the ground. But sometimes there are accidents. The oil spills and hurts the environment. People then need to work together to clean it up. The photo shows oil that washed up onto the shore. It gets all over the birds, fish, and sea creatures that live there.

1. Where is oil found?

____

____

2. What happens to the environment after an oil spill?

____

____

____

Think About It

Name: ______________________ **Date:** ______________

Directions: There was an oil spill in the ocean. Oil washed up onto the beach. Now, workers need to clean it up. Study the photo. Then, answer the questions.

1. Why do you think the cleanup crew is wearing masks and special clothing?

___

___

2. Does it look easy or hard to clean the beach? How do you know?

___

___

___

Name: ______________________ **Date:** ____________

Directions: Write a letter to the president of the United States. Tell the president how you feel about oil spills. Draw a picture to go with your letter.

Dear ______________________,

Sincerely,

Reading Maps

Name: ______________________________ **Date:** ______________

Directions: This map shows a region. There can be many communities in a region. Study the map. Then, answer the questions.

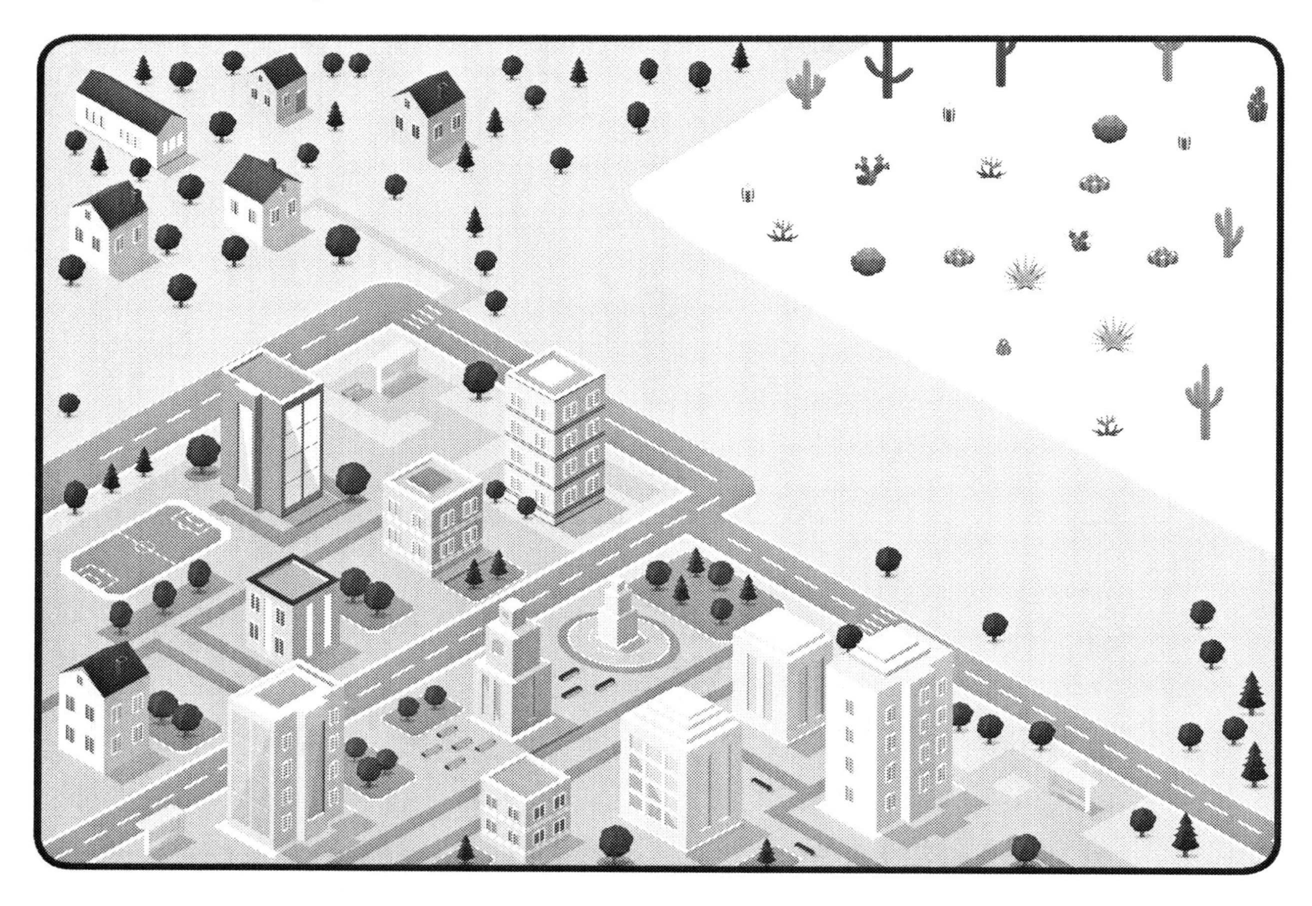

1. Circle the place with the most people living there.

2. Draw a box around the place with the fewest people living there.

3. Why might there be no people living there?

____

____

Name: ______________________ Date: ____________

Directions: Label the symbols. Use the word bank to help you. Then, tell a friend about the communities.

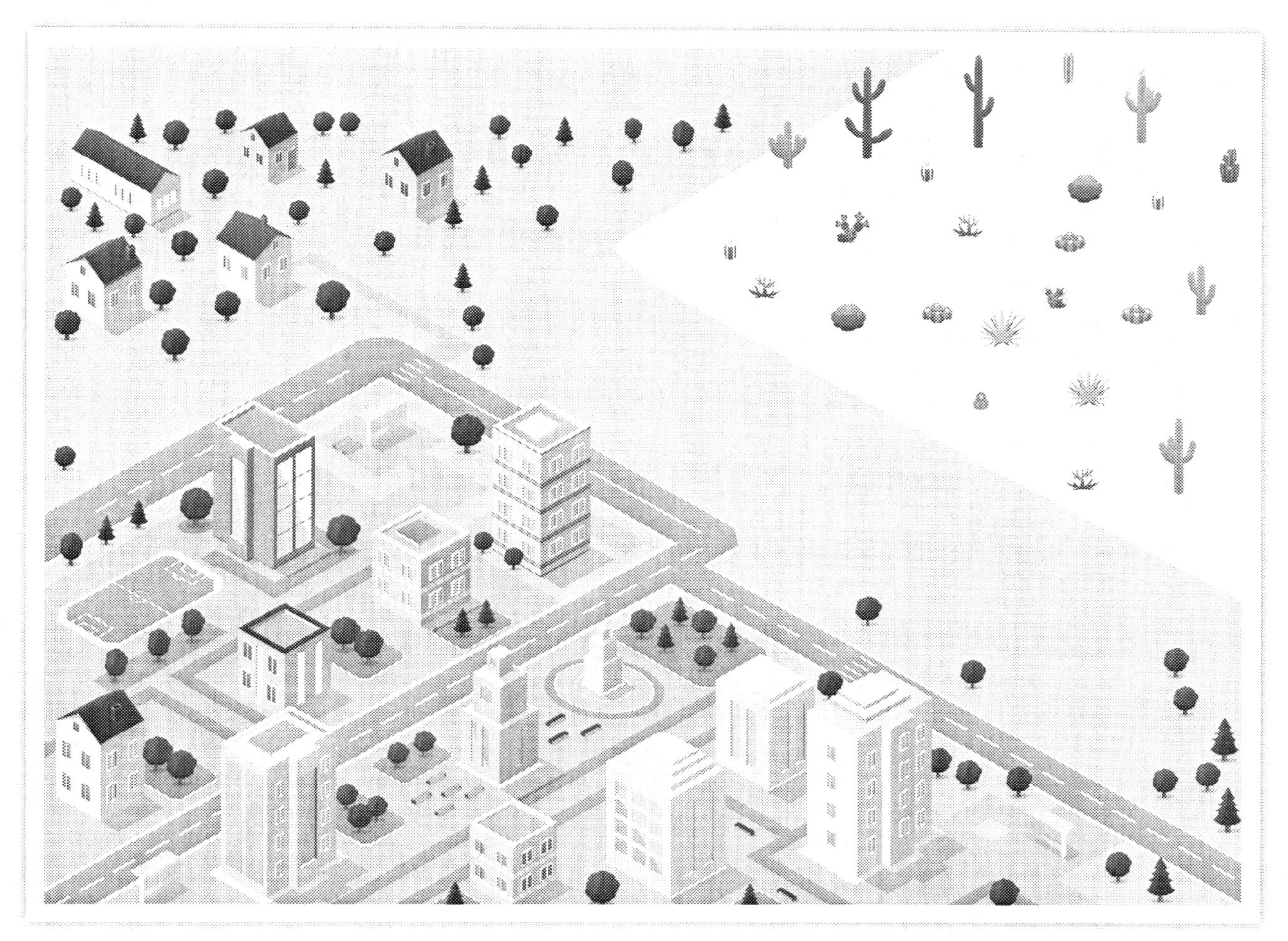

Word Bank		
no people	a few people	a lot of people

____

____

____

____

____

Name: ______________________ **Date:** ____________

Directions: Read the text. Study the photo. Then, answer the questions.

Life on the Coast

This is Australia. Most people there live on the coast. The middle of the continent is dry and gets little rain. But the coast gets a lot of rain. This photo shows a rain forest. It is right next to a beach. Explorers came to this beach to settle. It was near a lot of resources. But it was a hard life. There were floods. Pests brought diseases. And people were far from other towns and markets.

1. How is the middle of Australia different from the coast?

____

____

2. Why was life hard for the settlers?

____

____

Name: ______________________ **Date:** ______________

Directions: This map shows where people live in Australia. Study the map. Then, answer the questions.

People in Australia

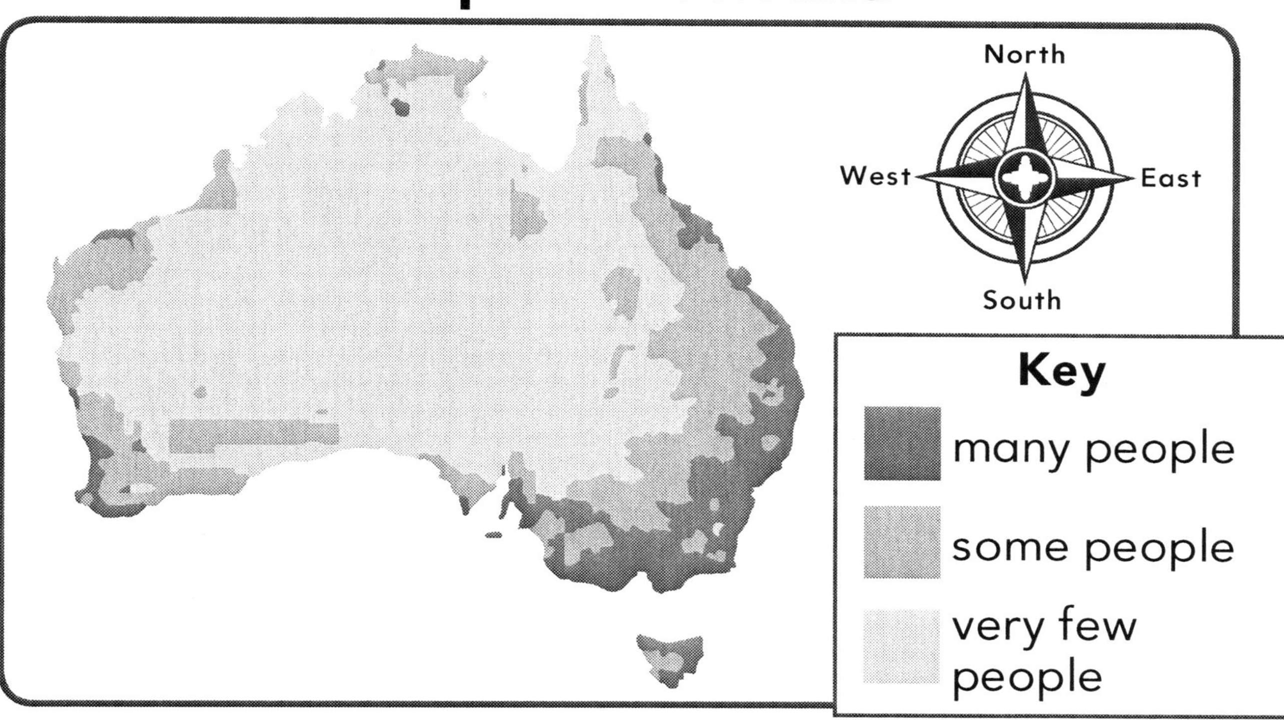

1. Where do most people live in Australia?

____

2. About how many people live in the center of Australia? How do you know?

____

____

3. Do more people live in eastern or western Australia? Tell a friend.

Name: ______________________________ **Date:** ______________

Directions: How would you feel about living in a place with very few people? Write and draw the positives and negatives.

Positives

Negatives

Name: ______________________ Date: ______________

Directions: This map shows two subway lines. Line 1 and Line 2 are paths the trains take. Study the map. Then, answer the questions.

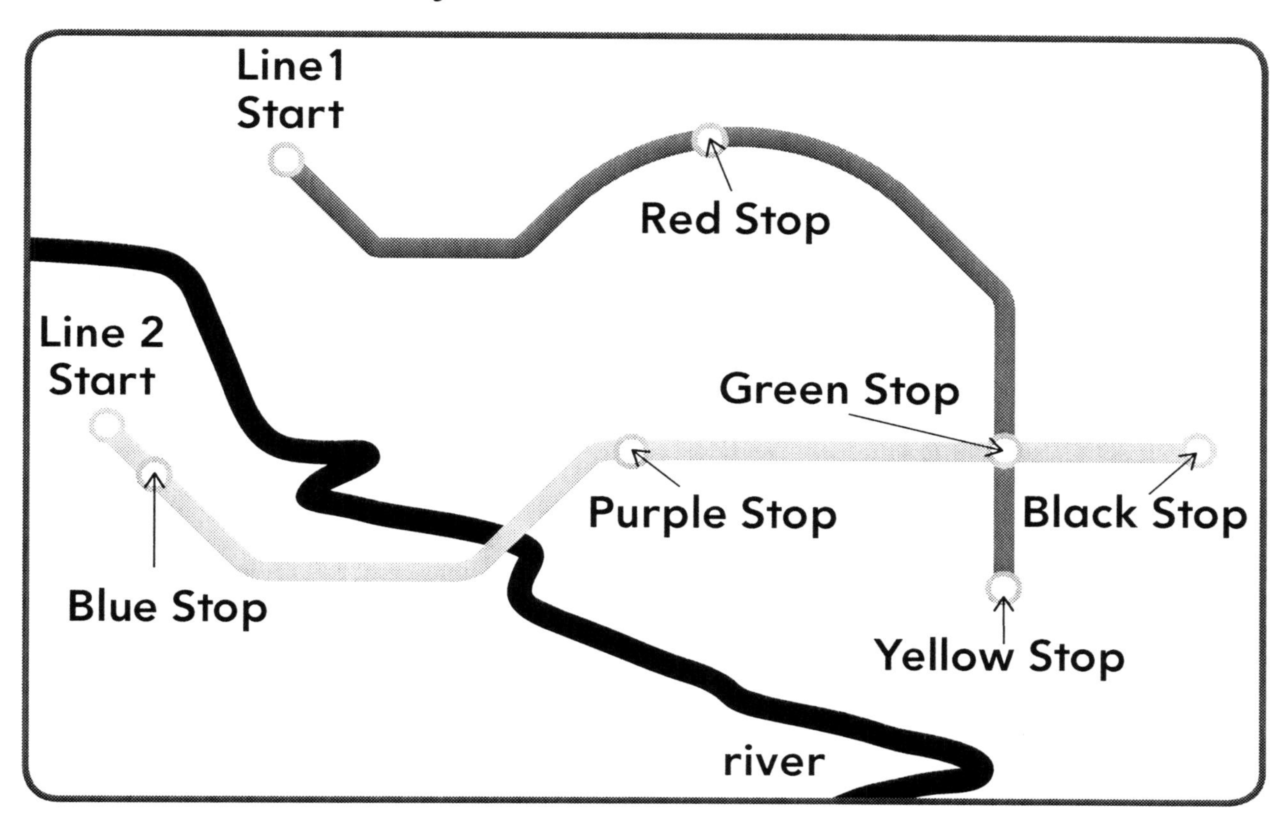

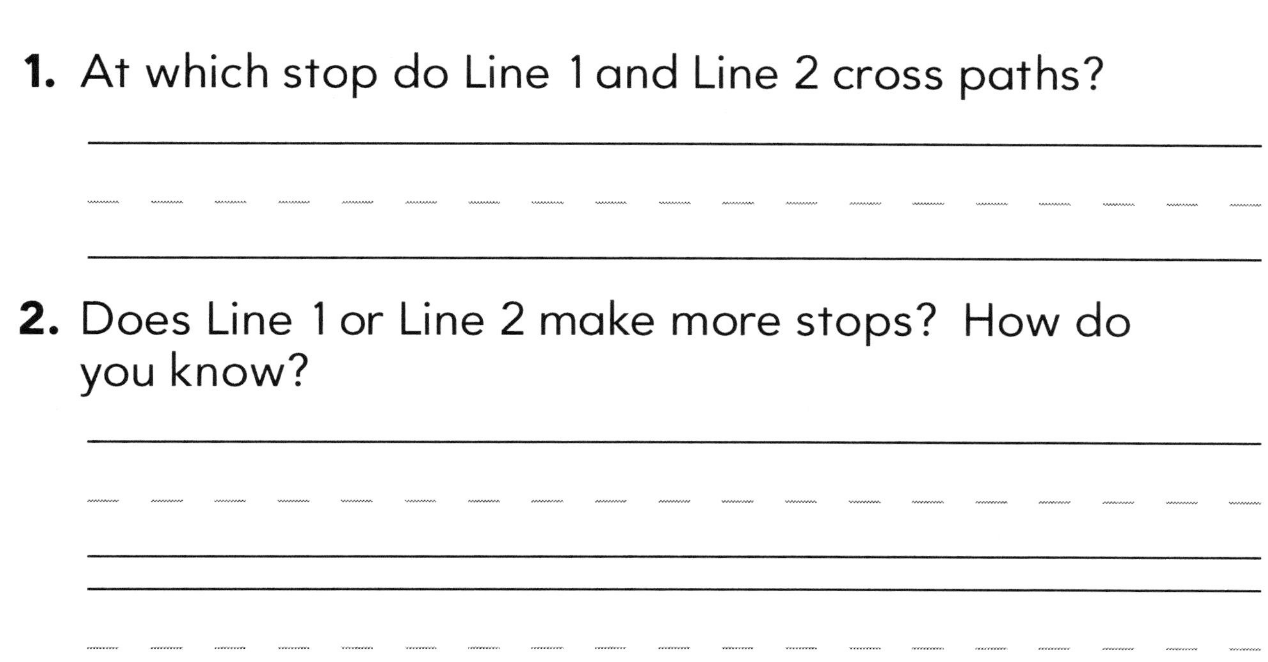

1. At which stop do Line 1 and Line 2 cross paths?

2. Does Line 1 or Line 2 make more stops? How do you know?

Name: ______________________ Date: ____________

Directions: Follow the steps.

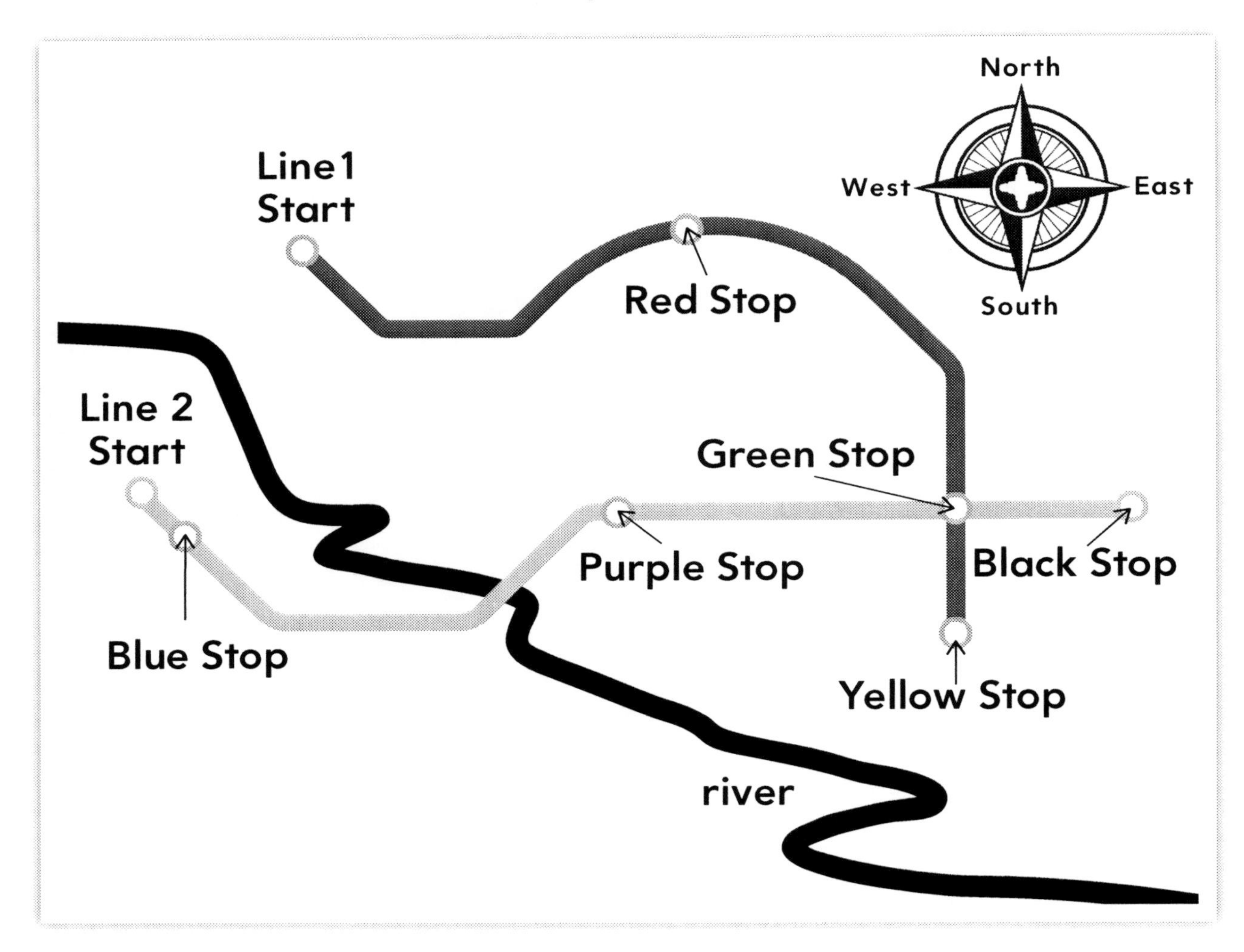

1. Circle where a subway line crosses the river.
2. Draw a box around the stop that is west of the river.
3. Draw a triangle by the stop that is farthest south.
4. Draw a star where both lines meet.
5. How could you go from the Line 1 Start to the Black Stop? Tell a friend.

Name: ______________________ Date: ____________

Directions: Read the text, and study the photo. Then, answer the questions.

The Subway

A subway is like a train. It goes underground. The railcars roll on tracks. The tracks are in long tunnels. They are built under buildings and streets. A subway makes many stops. Each stop is a different place.

The subway is a fast way for people to move around the city. Driving can take a lot of time. Cities are busy. There is a lot of traffic. Riding on the subway can save people time.

1. Why do some people ride the subway instead of driving?

____

____

2. How is a subway different from a regular train?

____

____

Think About It

Name: ______________________ **Date:** ____________

Directions: This bar graph shows how many people ride the subway before noon. Use the graph to answer the questions.

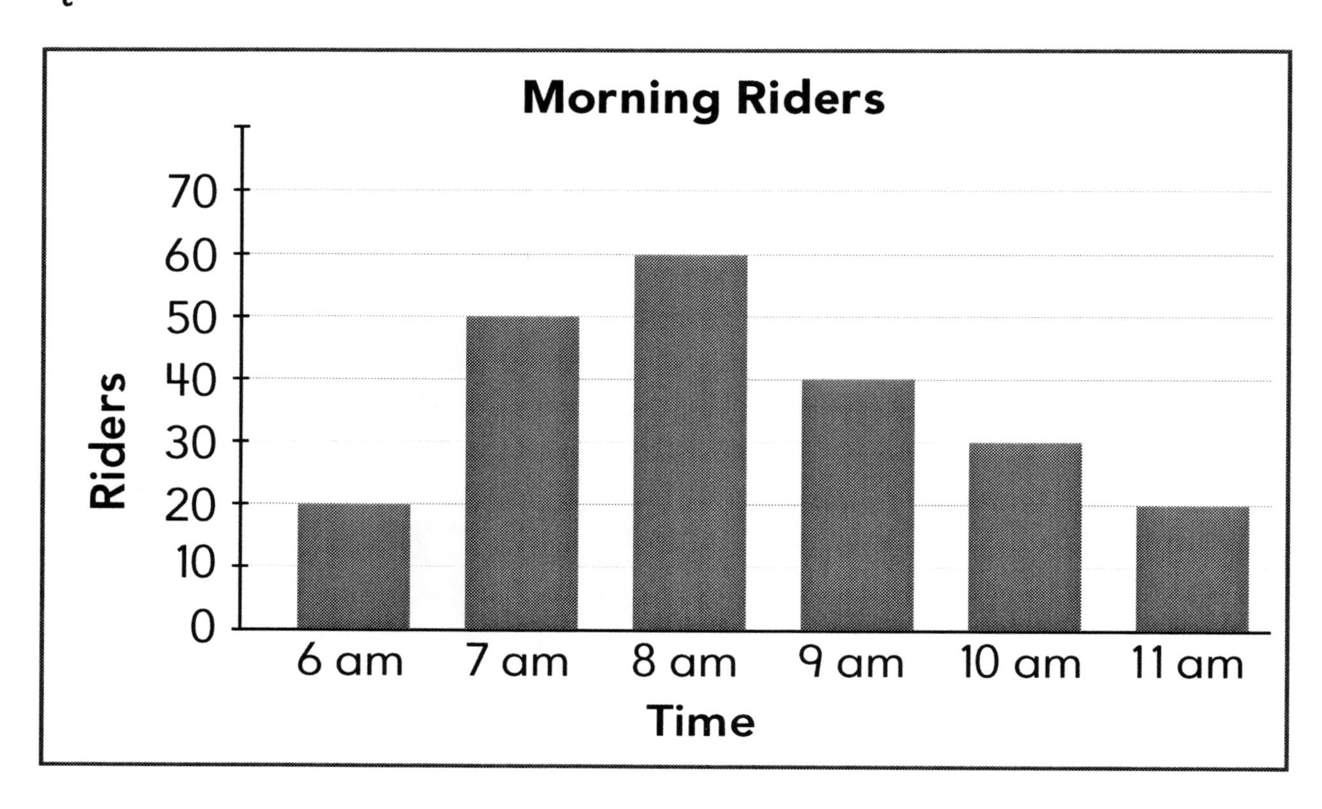

1. At what times are there fewer than 30 riders?

____

2. At what time do you think most people are going to work? How do you know?

____

____

Name: ______________________ **Date:** ______________

Directions: Would you would rather travel around a large city in a car or on the subway? Why? Draw and write your answer.

Name: ______________________ Date: ____________

Directions: This map shows the Nile River. Study the map. Then, answer the questions.

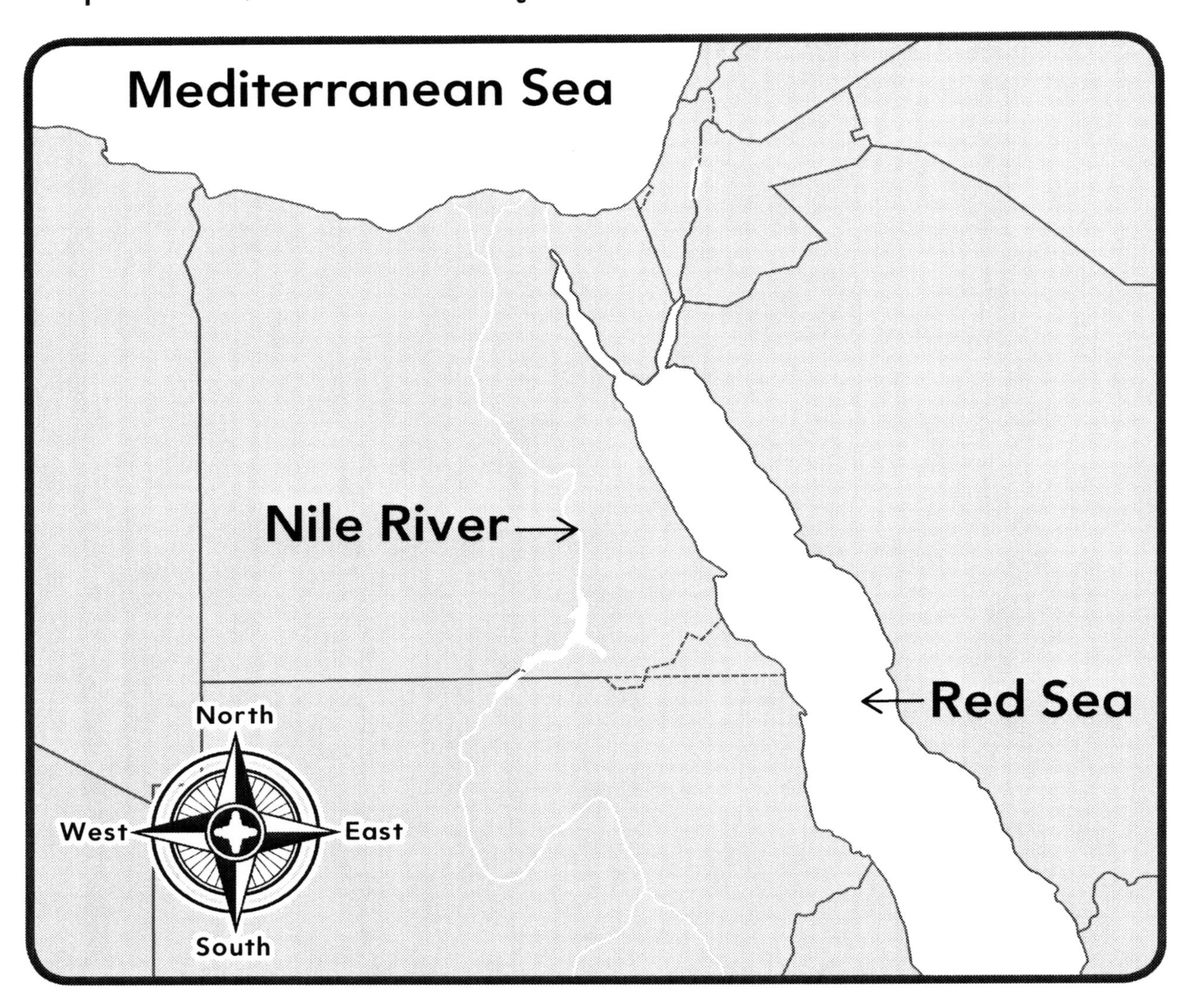

1. Trace the Nile River in blue.

2. Describe the shape and size of the river.

Name: ______________________ **Date:** ______________

Directions: Follow the steps.

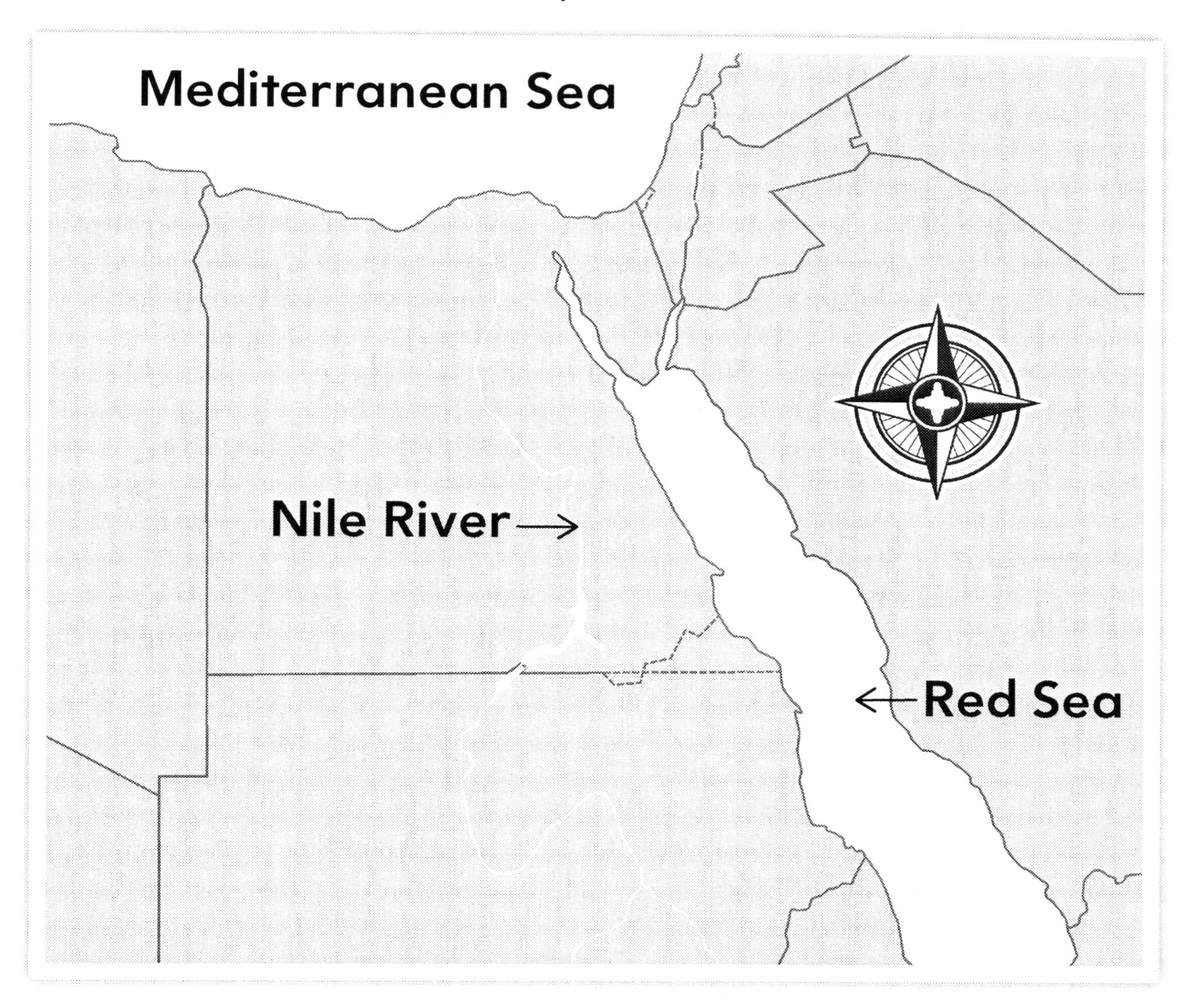

1. Circle where the river touches the sea.

2. Complete the compass rose.

3. Color all the bodies of water blue.

Read About It

Name: ______________________ **Date:** ____________

Directions: Read the text. Study the photo. Then, answer the questions.

Rivers

Many large cities are built right next to a river. Long ago, people needed to live close to the river. It gave them resources they needed to live. It gave them food and fresh water. They drank the water. They used it to keep clean. They used it to water their crops. A river also helped people get from place to place. They traveled up and down the river by boat.

1. What are some resources found in a river?

____

____

2. Why are some cities built next to a river?

____

____

Name: ______________________ Date: ______________

Directions: Compare three rivers from around the world. Then, answer the questions.

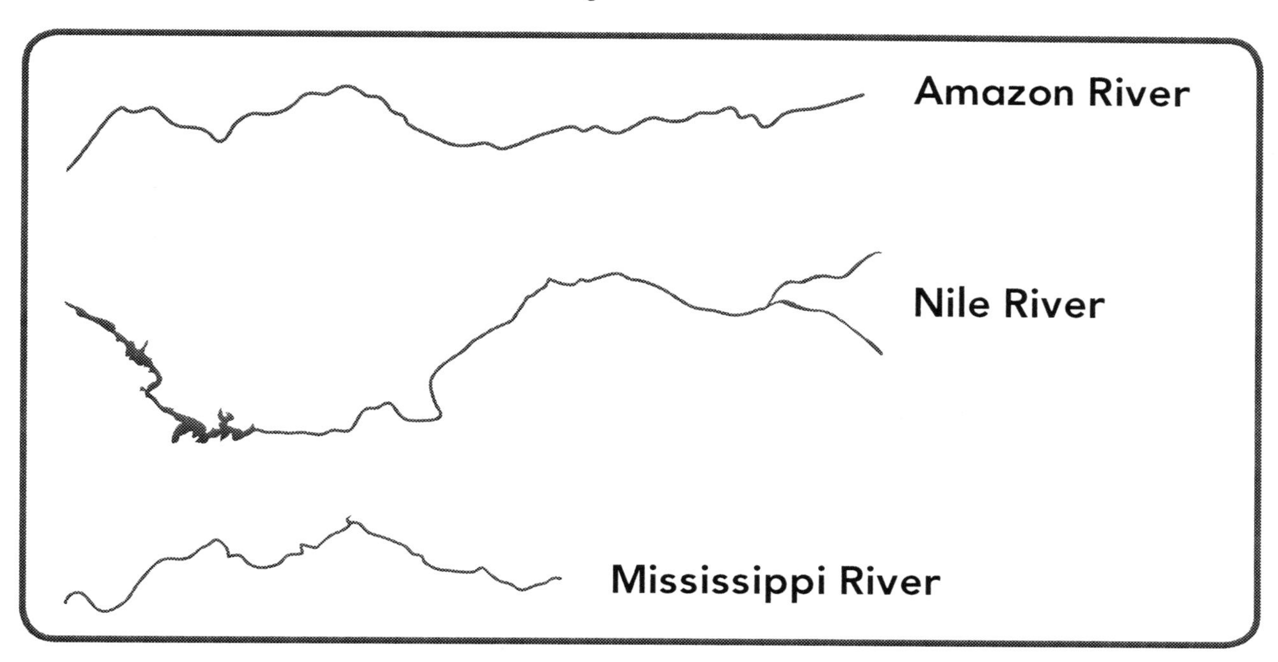

1. Which two rivers are almost the same length?

2. Why might the Nile River look thicker on one end?

Think About It

Name: ______________________________ **Date:** ______________

Directions: Draw four ways you use water.

Name: ______________________ **Date:** ____________

Directions: This is a map of a suburban community. Study the map. Then, answer the questions.

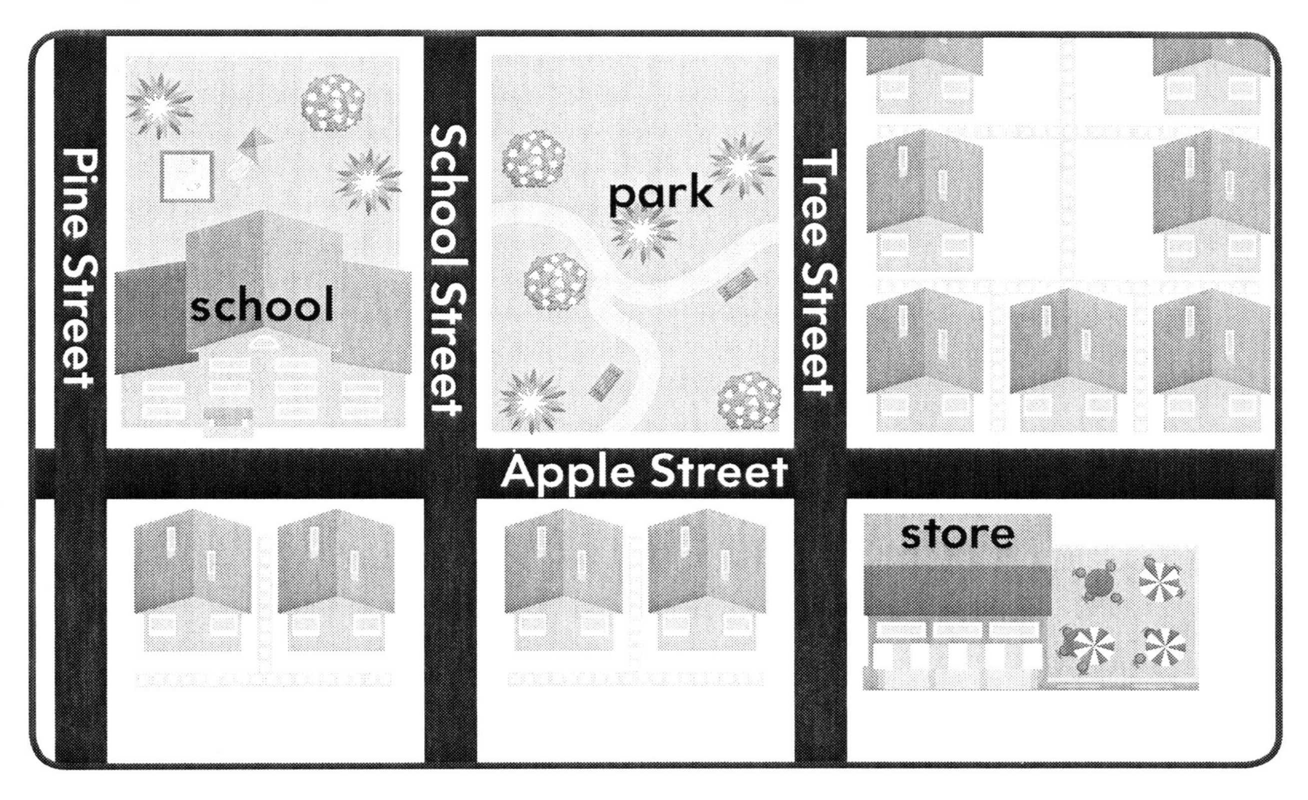

1. Which street does the school face?

____

2. What is across from the school?

____

3. Is the school closer to the park or the store?

____

Name: ______________________________ **Date:** ____________

Directions: Draw a student in the park. Write directions telling how that student can get to the store. Use the cardinal directions to help you.

Pine Street
school
School Street
park
Tree Street
North
West
East
South
Apple Street
store

____

____

____

____

Name: ______________________ Date: ____________

Directions: Read the text. Study the photo. Then, answer the questions.

Read About It

Rows of Homes

Suburbs are often found outside cities. They are not too crowded. Most of the land there is used for houses. It has some businesses a short drive away. The homes have grassy front and backyards. People can work in the suburbs. But many people still work in the city. They commute to get there. The drive is not too long.

1. How are the suburbs different from a city?

____

____

2. How can people live in the suburbs and work in the city?

____

____

Think About It

Name: ______________________________ **Date:** ________________

Directions: Katie is trying to pick a place to have a birthday party. Study the table. Then, answer the questions.

Place	Star Arcade	Gold Theater	Family Fun Center	Mini-Golf Park
Miles	4 miles	1 mile	2 miles	3 miles

1. Which place is the farthest from Katie's house?

____

2. If Katie wants the party to be close to her house, where should she have it?

____

3. Why might someone want to have a party close to his or her home?

____

____

4. Tell a friend how you used the chart to answer the questions.

Name: ______________________ Date: __________

Directions: Do you live in a rural, suburban, or urban community? Draw and write about your community.

Name: ______________________________ **Date:** ______________

Directions: The truck takes goods to the farmers' market. Study the map. Answer the questions. Then, color the map.

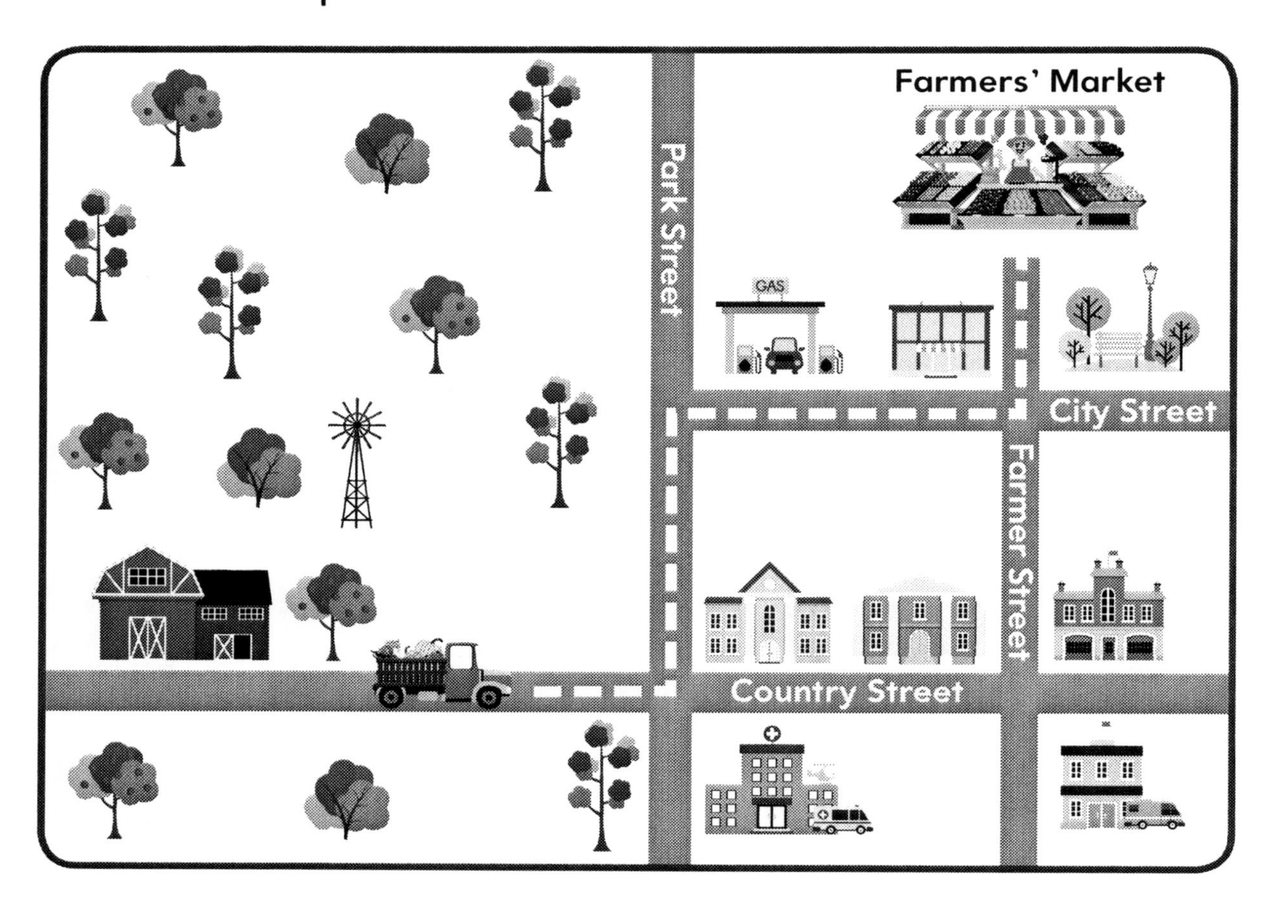

1. Circle the farmers' market.

2. Name two other places on the map.

____

____

____

Name: ______________________ **Date:** ____________

Directions: Write directions the farmer will take to the farmers' market. Use the street names and the words *left* and *right* in your directions.

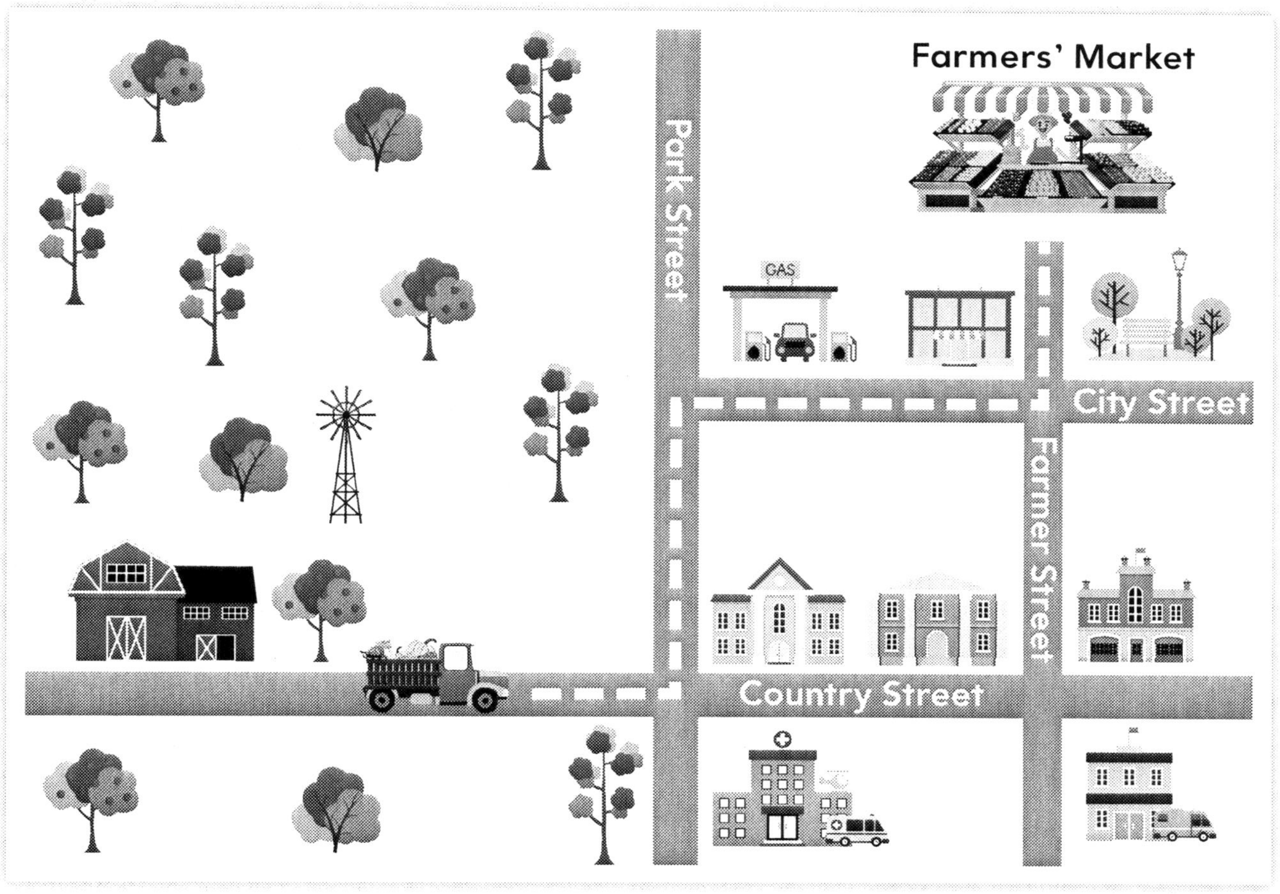

Farmers' Market

GAS

Park Street

City Street

Farmer Street

Country Street

Name: ______________________ **Date:** ____________

Read About It

Directions: Read the text. Study the photo. Then, answer the questions.

Pike Place Market

Pike Place is in Seattle, Washington. Goods from local farms are sold here. Farmers sell fruits and vegetables. They sell flowers, nuts, honey, and jam. It is also a place for artists to come. They bring their crafts. They set up booths to show their art. Pike Place Market is an open-air market. But it has a roof. So people come to shop—rain or shine.

1. Where do the goods at a farmers' market come from?

____

____

2. Why do artists come to the farmers' market?

____

____

Name: ______________________ Date: ____________

Directions: This graph shows the goods sold at a farmers' market one morning. Study the graph. Then, answer the questions.

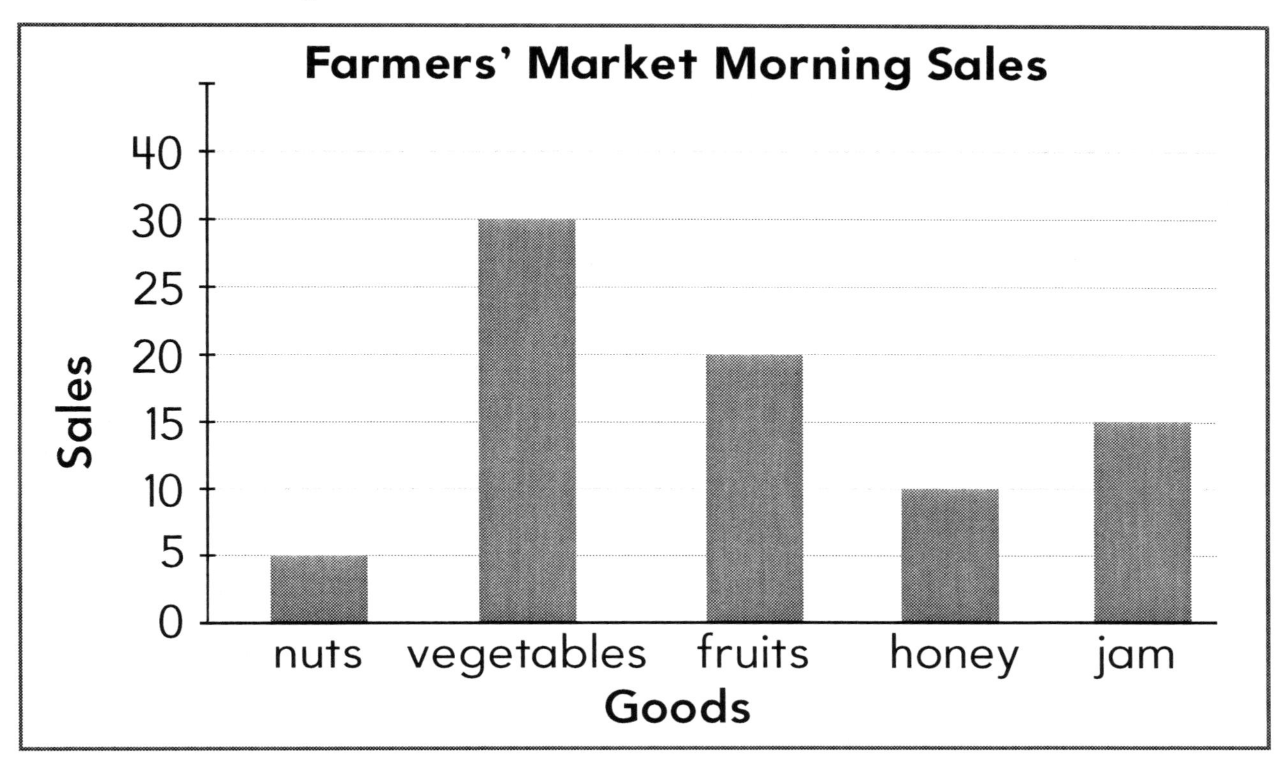

1. Which item sold the most?

____

2. Which item sold the least?

____

3. How many jars of jam were sold?

____

Name: ______________________________ Date: ______________

Directions: Think about a store where you can buy fruits and vegetables. Compare and contrast that place with a farmers' market.

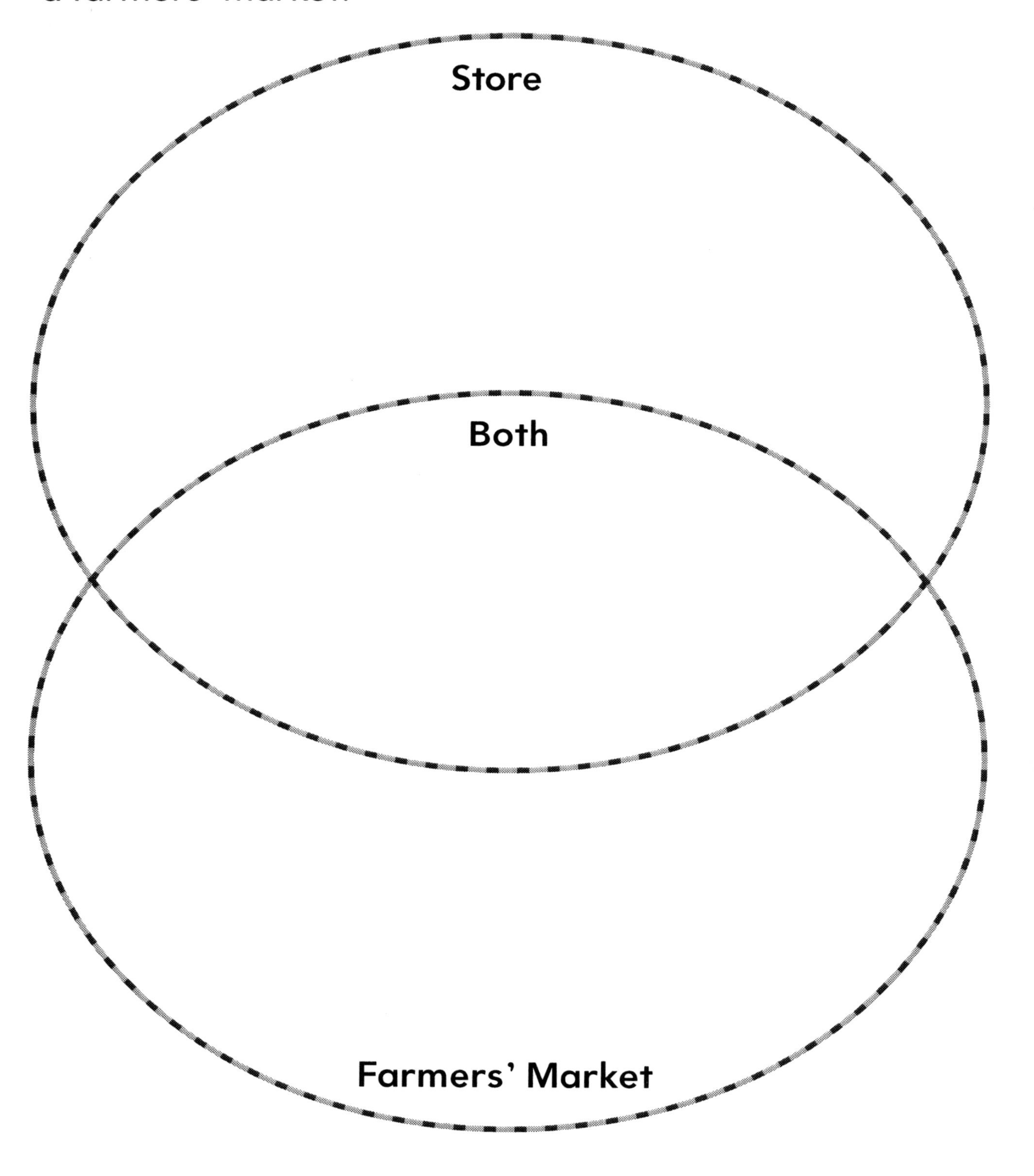

Name: ______________________ **Date:** ______________

Directions: Imagine you are moving to a new class. Follow the steps to find your seat.

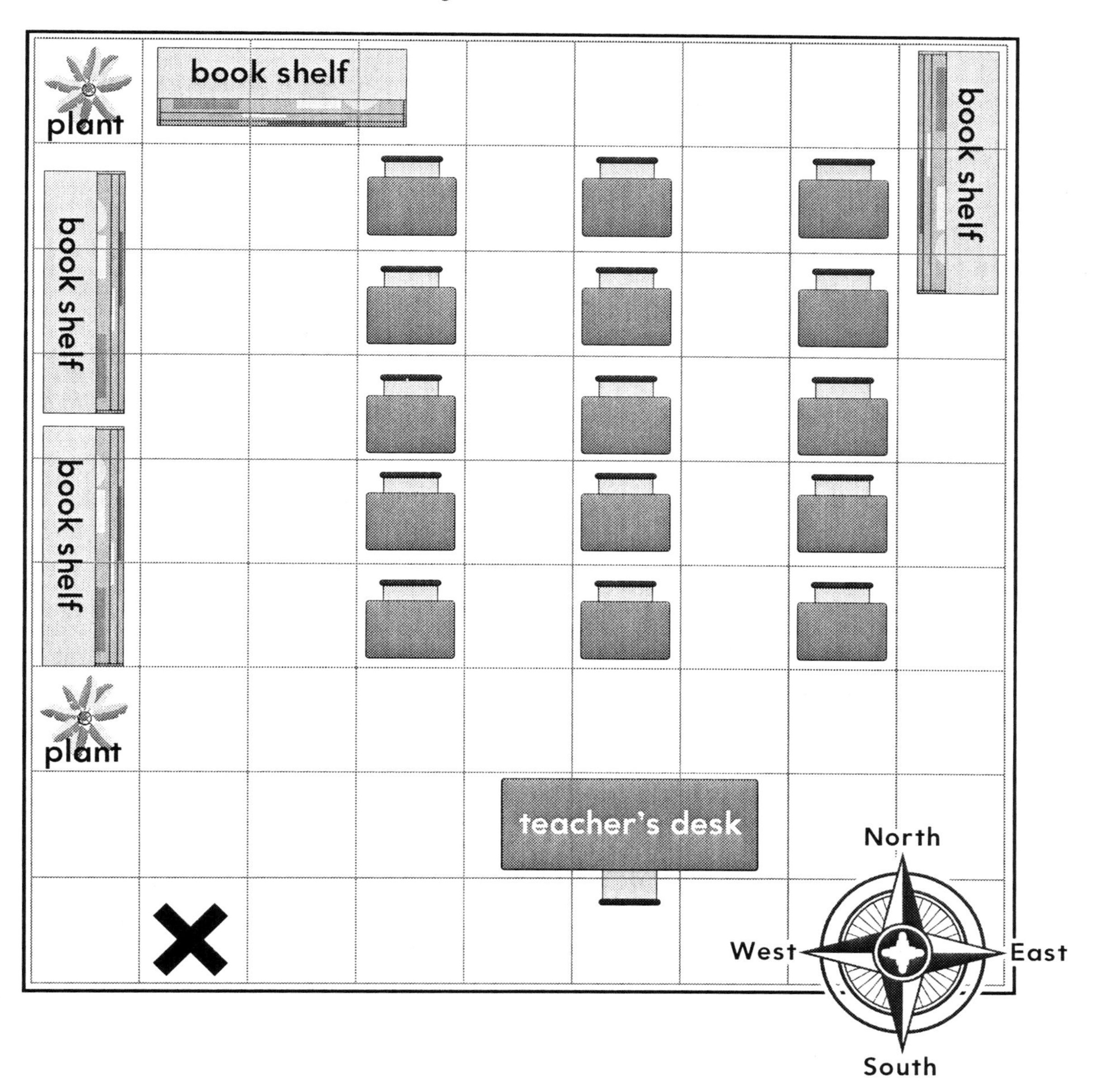

1. Start at the X. Walk four squares north.
2. Walk six squares east. Circle the desk you are next to.
3. Write your name next to that desk.
4. Describe to a friend where this seat is.

Name: ______________________________ Date: ______________

Directions: Draw and label a map of your classroom.

Creating Maps

Name: ______________________ Date: ______________

Directions: Read the text. Study the photo. Then, answer the questions.

School Around the World

School is not the same for every kid. How do you get to school? In Uganda, some kids have to cross a lake in a canoe. In Australia, some kids take the train. Kids in different countries also learn about different things. Have you thought about who your friends are? In some countries, boys and girls do not go to school together. Boys go to school with boys. And girls go to school with girls. What is school like where you live?

1. Study the photo. What do all the students have in common?

____

____

2. How is school different around the world?

____

____

Think About It

Name: ______________________ **Date:** ____________

Directions: This graph shows the number of hours different first graders spend in school each day. Study the table. Then, answer the questions.

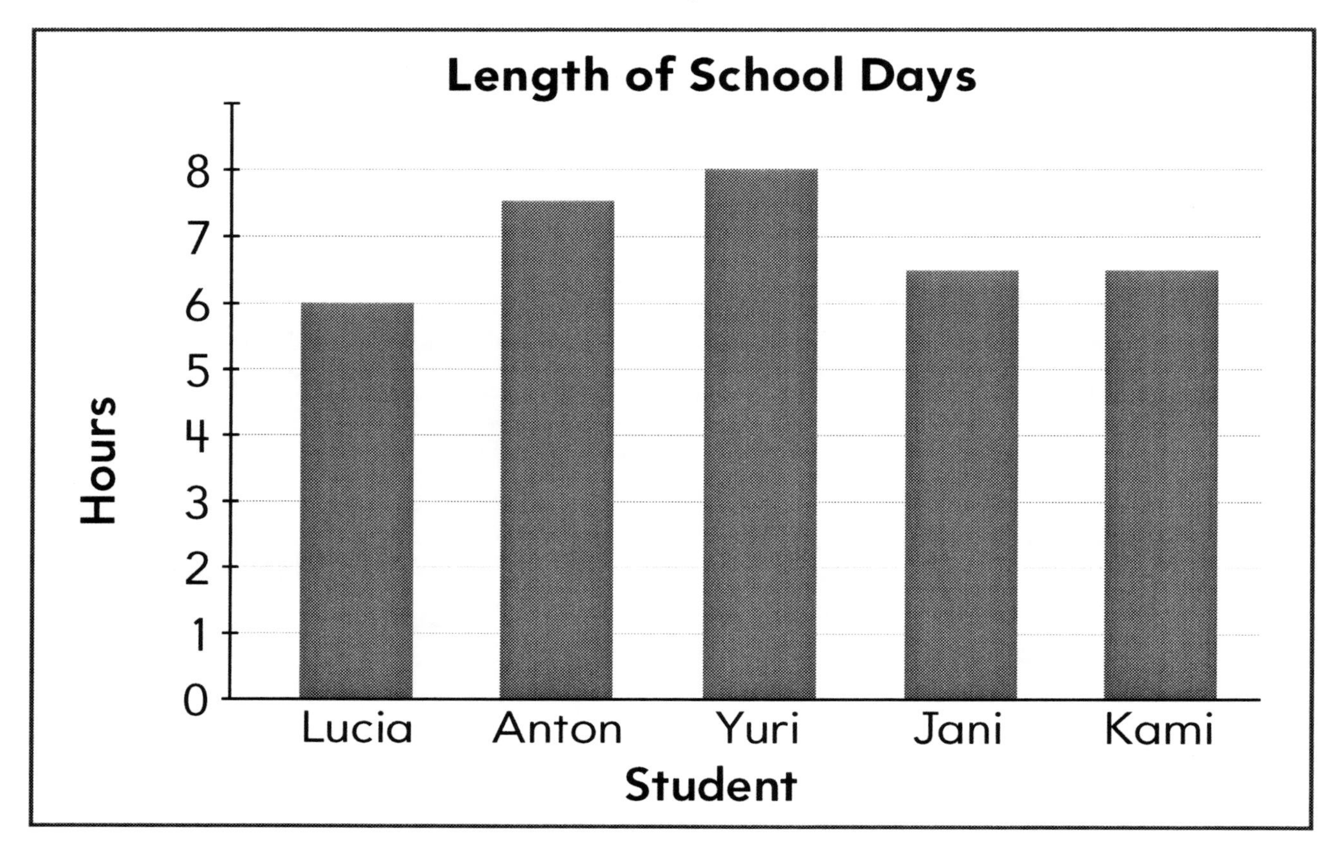

1. Which student has the shortest school day?

____

2. Which student has the longest school day?

____

3. Tell a friend how your school day compares to the ones in the graph.

Name: ______________________________ Date: ____________

Directions: Compare your school to a school in Japan.

My School	School in Japan
	wear uniforms
	go to school for seven hours each day
	have 40 days of summer vacation
	get two recesses each day

Name: ______________________ **Date:** ____________

Directions: This is a map of a backyard. Study the map. Then, answer the questions.

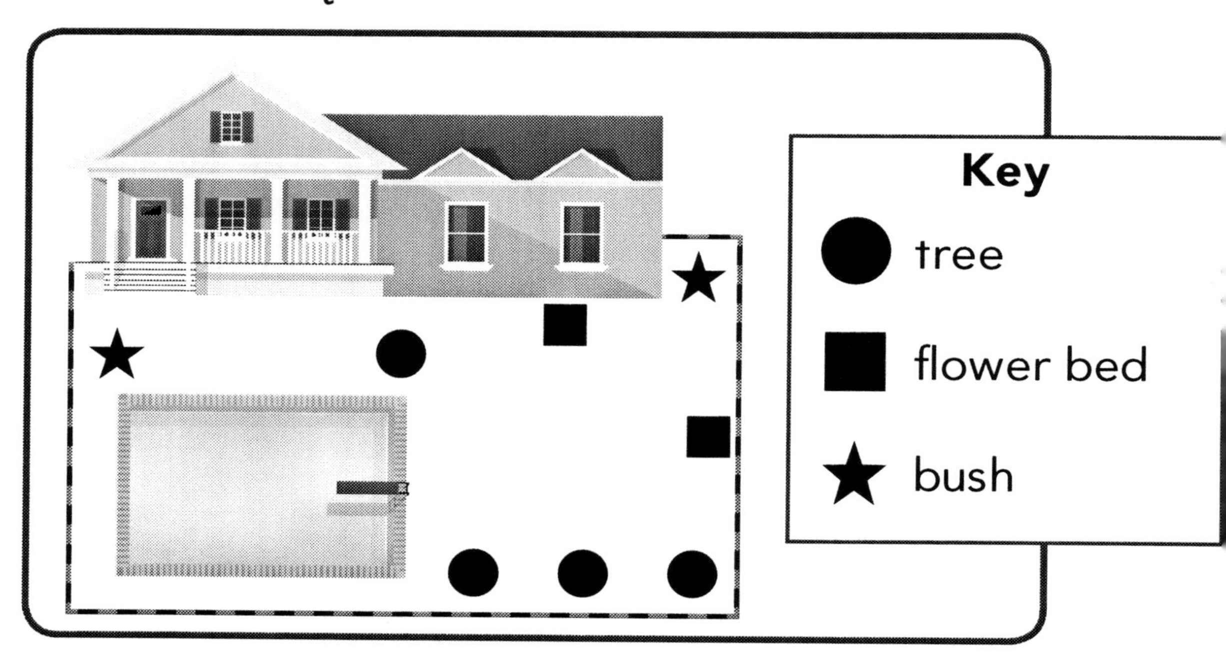

1. Will there be shade next to the pool? How do you know?

____

____

2. Do you think there will be enough flowers? Why or why not?

____

____

Name: ______________________________ Date: ____________

Directions: Draw a map of a backyard garden. Use the map key to help you.

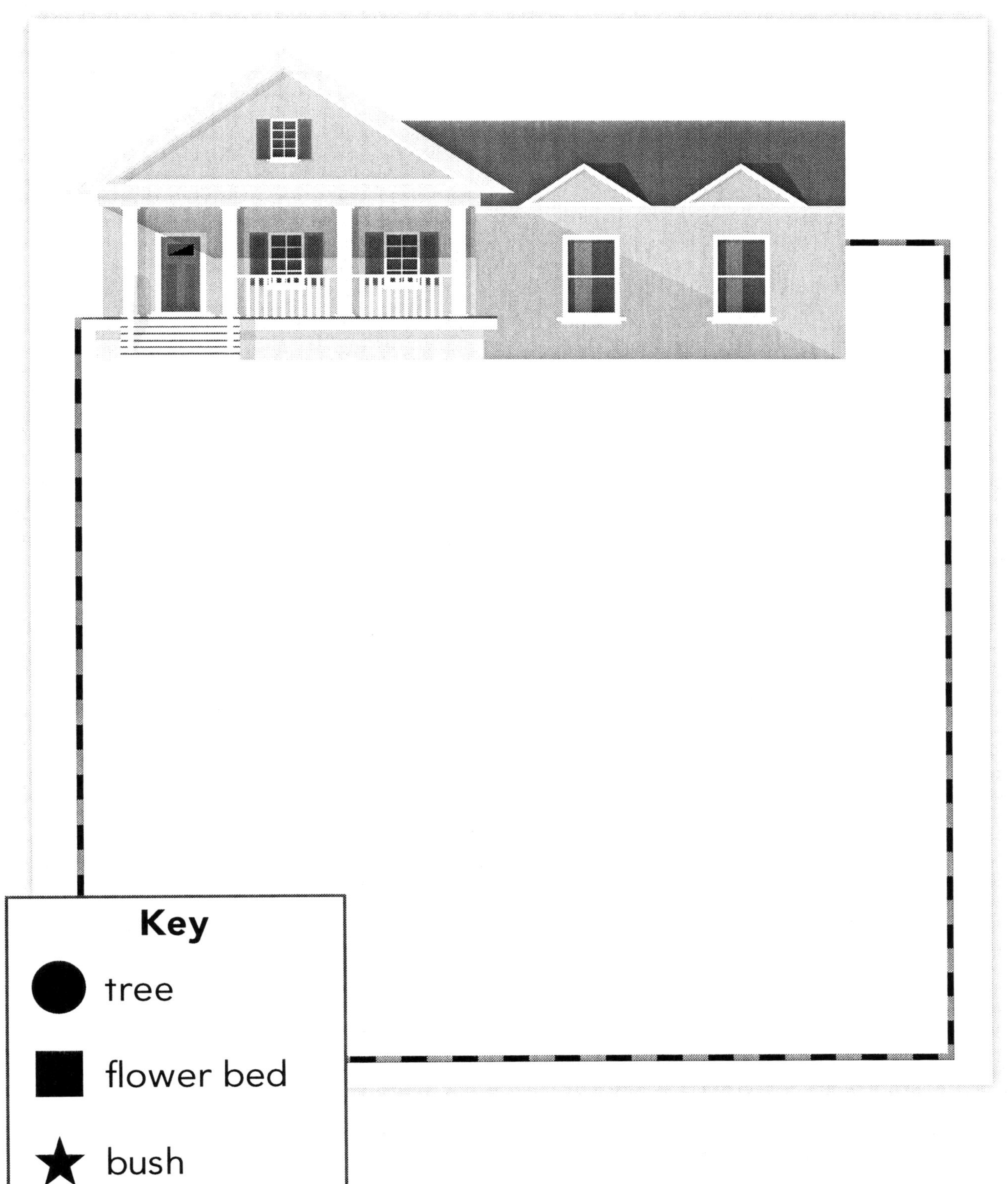

Key

● tree

■ flower bed

★ bush

Read About It

Name: ________________________ **Date:** ____________

Directions: Read the text. Study the photo. Then, answer the questions.

Parks and Gardens

Parks and gardens are special places. The picture shows Butchart Gardens. It is known for its flower beds. A garden like that is planned. Landscape architects plan where to plant bushes, trees, and flowers. They make maps to show where each type of plant goes. It helps them know what it will look like before they start planting.

1. How can you tell that the garden in the photo was planned?

____

____

2. What does a landscape architect do?

____

____

Name: ______________________ Date: ____________

Directions: Study the diagram. Then, answer the questions.

City Map	Both	Garden Map
large area	show paths	small area
shows a lot of places	show where things are	shows one place
shows streets		shows the types of plants

1. How are city and garden maps different?

____

____

2. What is one thing city and garden maps have in common?

____

3. Add a detail to each part of the Venn diagram.

Name: ______________________ Date: ____________

Directions: Draw and write about a park or garden in your community.

Name: ______________________ **Date:** ____________

Directions: This map shows the stops during a tour at the zoo. Study the map. Then, answer the questions.

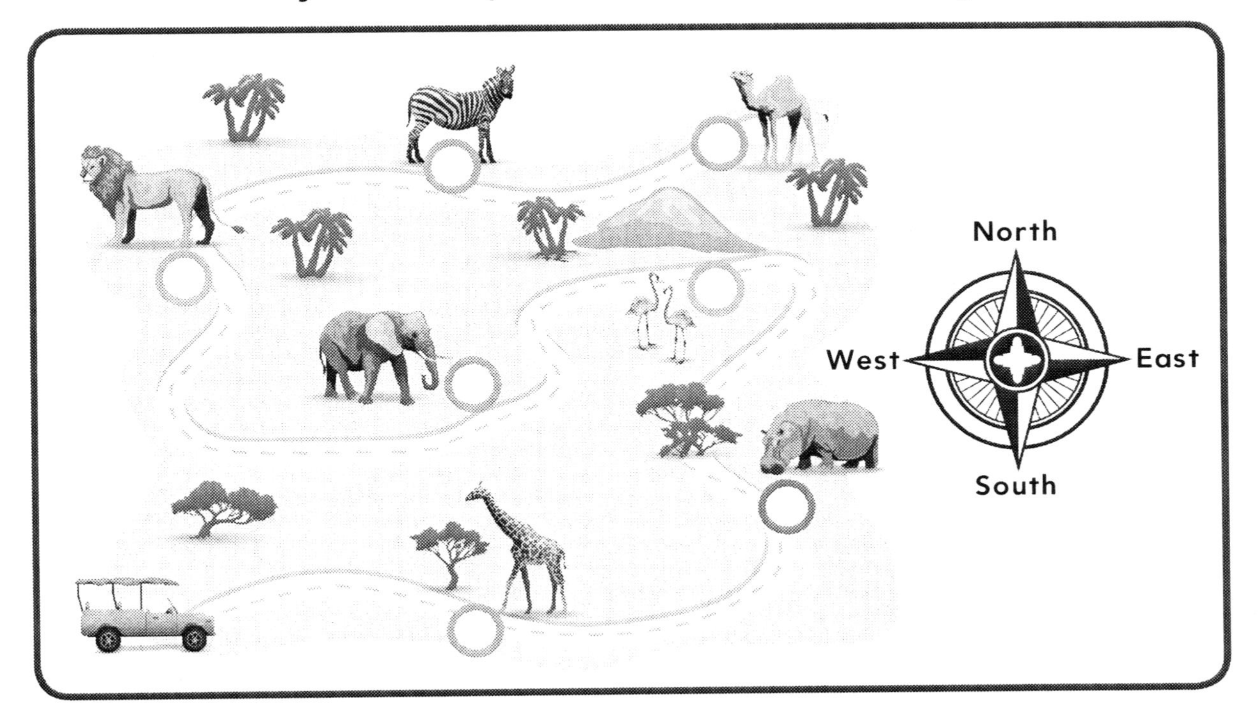

1. What animal will you see at the first stop?

2. How many stops will you make before you get to the lion?

3. The tour begins in the south part of the map. What part of the map does the tour end in?

Name: ______________________________ **Date:** ____________

Directions: Use the compass rose. Which direction should the tour guide drive next? Label each circle on the path with one of the cardinal directions.

Name: ______________________________ Date: ______________

Directions: Read the text. Study the photo. Then, answer the questions.

Life on the Savanna

The savanna is grassy and flat. It is a great home for many animals. That brings tourists. Those are people who visit a place. They want to go on safaris. Tour guides know where to take them. But the weather is getting drier there. And the number of people who live there is growing, too. So the animals are losing their homes. Now, some animals are endangered. That means there are not a lot of them left.

1. Why do people go on safaris?

____

____

2. What is one reason the animals are endangered?

____

____

Think About It

Name: ______________________ **Date:** ____________

Directions: Draw lions in one part of the picture. Then, draw people in the other three parts. Study your drawing. Then, answer the questions.

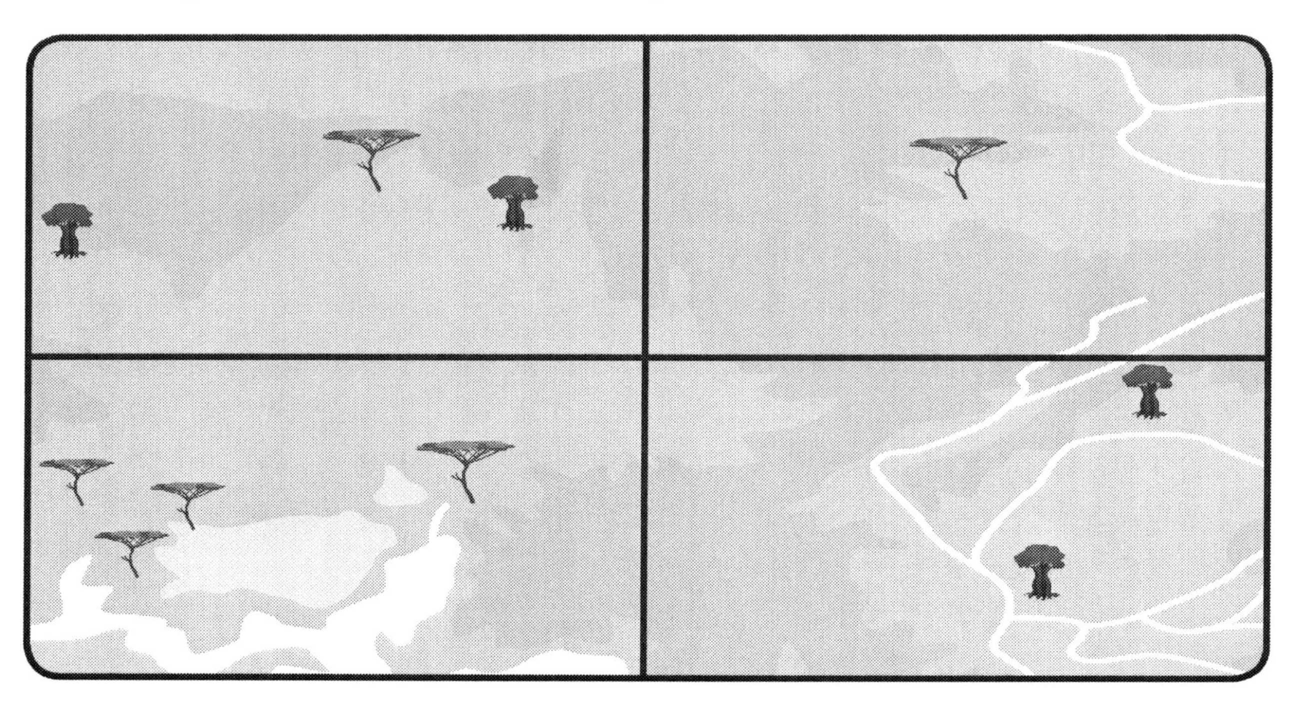

1. The lions used to have all four parts of the land. How much of the land do lions live on now?

____

2. What does this picture tell us about what people have done to the land?

____

____

Name: ______________________ Date: ____________

Directions: People need places to live. Pretend you live on the savanna. Draw and write about sharing the lions' land.

Name: ______________________________ **Date:** ______________

Directions: A globe is a model of Earth. Study the globe. Answer the questions.

1. Label water and land.

2. How is a globe similar to a map?

____

____

Name: ______________________________ **Date:** ______________

Directions: Follow the steps to complete the globe.

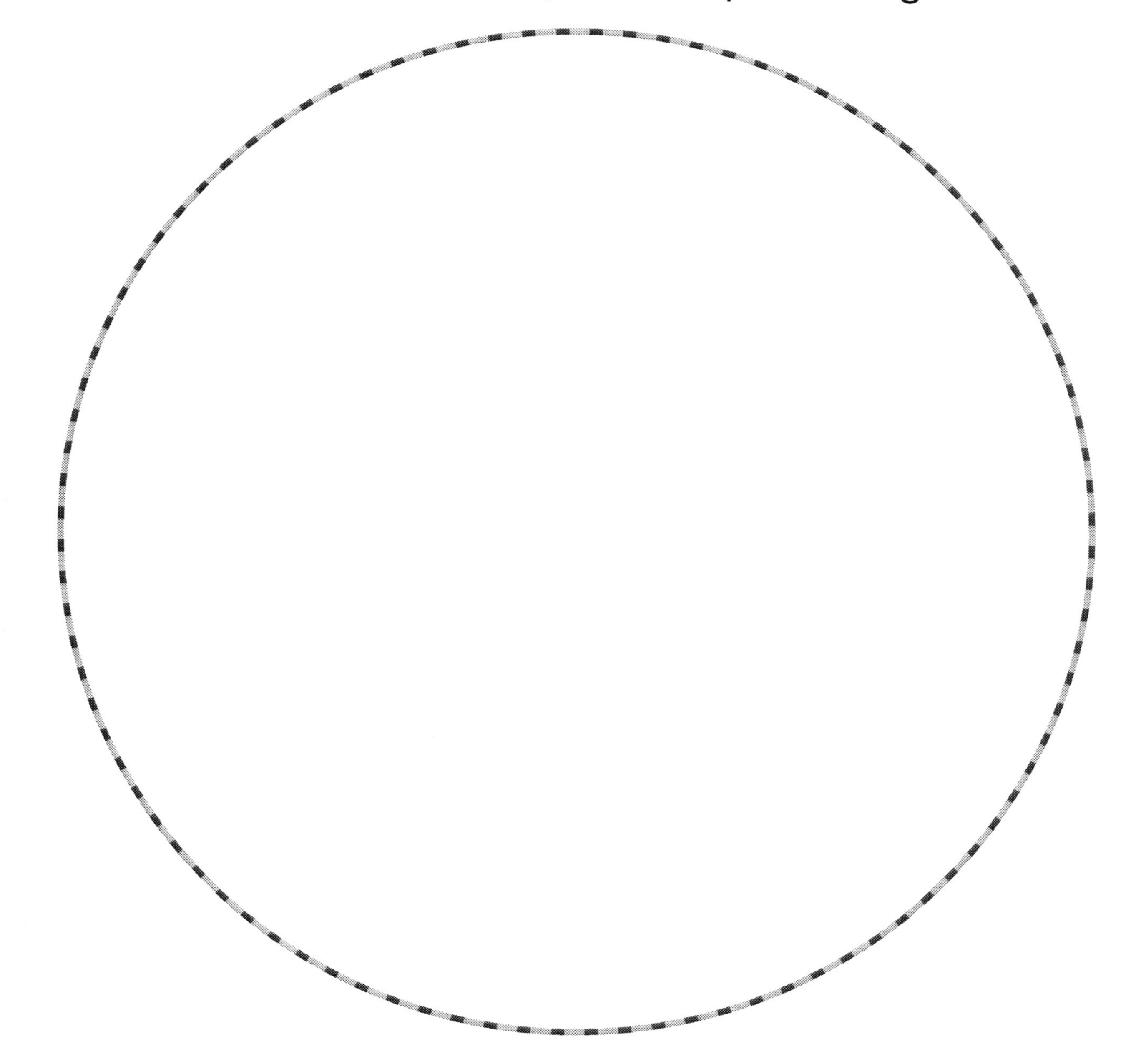

1. Draw some land. Color the land green and the water blue.

2. Draw a star to show where people can live.

3. Write a made-up name for that place next to the star.

Name: ______________________ Date: ____________

Directions: Read the text. Study the photo. Then, answer the questions.

Read About It

The Coldest City

People who live in Yakutsk, Russia, face cold weather for much of the year. They dress in layers of warm clothes. They wear parkas. Those are thick, warm coats. They wear hats called ushanka. It means "hat with ear-flaps" in Russian. It is a fur-lined hat that covers a person's ears. People in this city know how to stay warm!

1. Describe a ushanka.

____

____

2. Describe what a cold, snowy day looks like in Yakutsk.

____

____

Name: ______________________ Date: ____________

Directions: Kate looked up the weather for her city. The graph shows the average high temperature for every month. Use the graph to answer the questions.

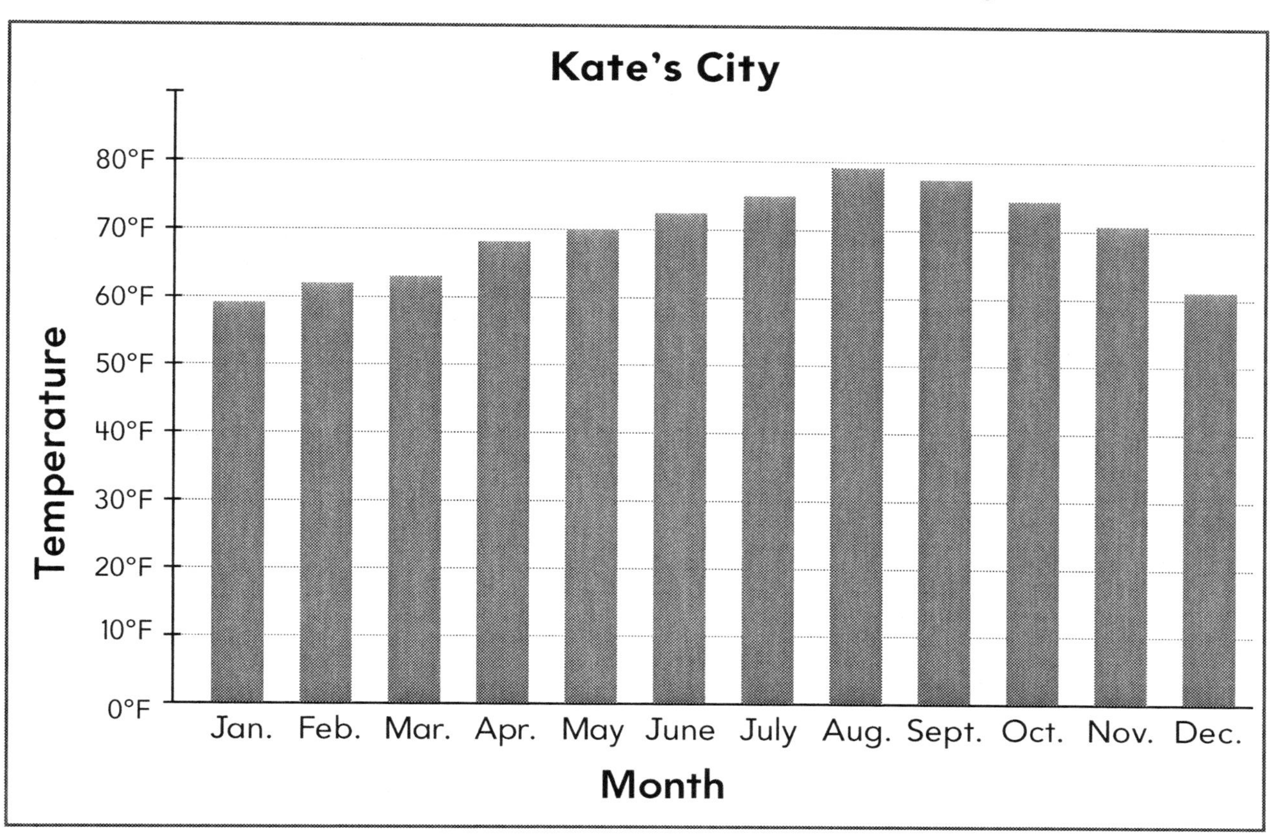

1. Which is the warmest month where Kate lives?

____

2. Which are the coldest months?

____

____

Name: ______________________ **Date:** ____________

Directions: How are the clothes people wear in Russia during winter similar to your clothes? How are they different? Draw and write your answer.

Name: ______________________ Date: ______________

Directions: An island is land that is surrounded by water. The map shows a chain of islands. Study the map. Then, answer the questions.

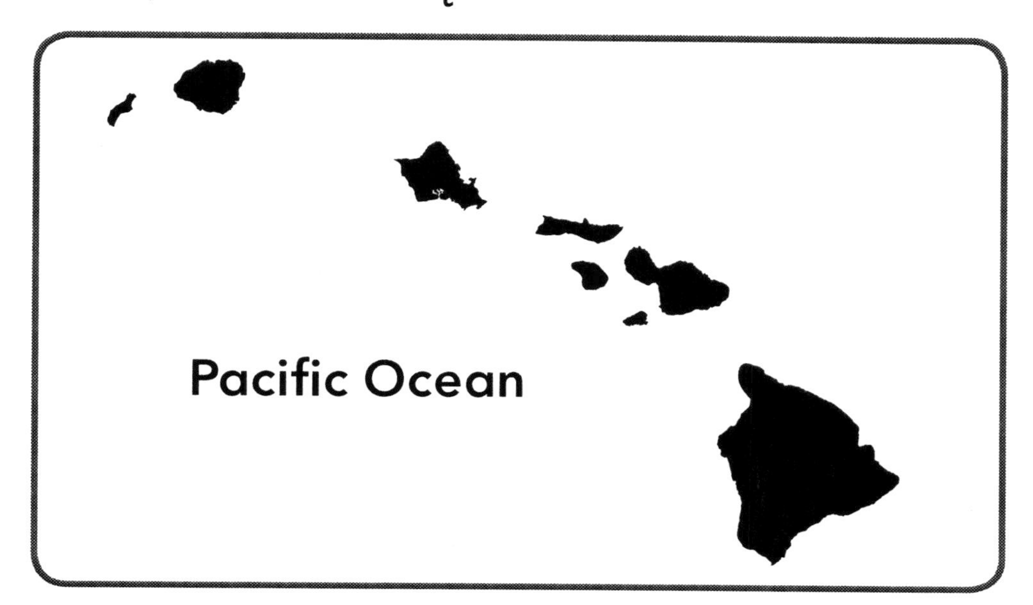

1. How many islands are part of the chain?

____

2. Which ocean surrounds the islands?

____

3. Why do you think they call it a "chain of islands"?

____

____

Creating Maps

Name: ______________________ **Date:** ______________

Directions: Follow the steps.

Hawai'ian Islands

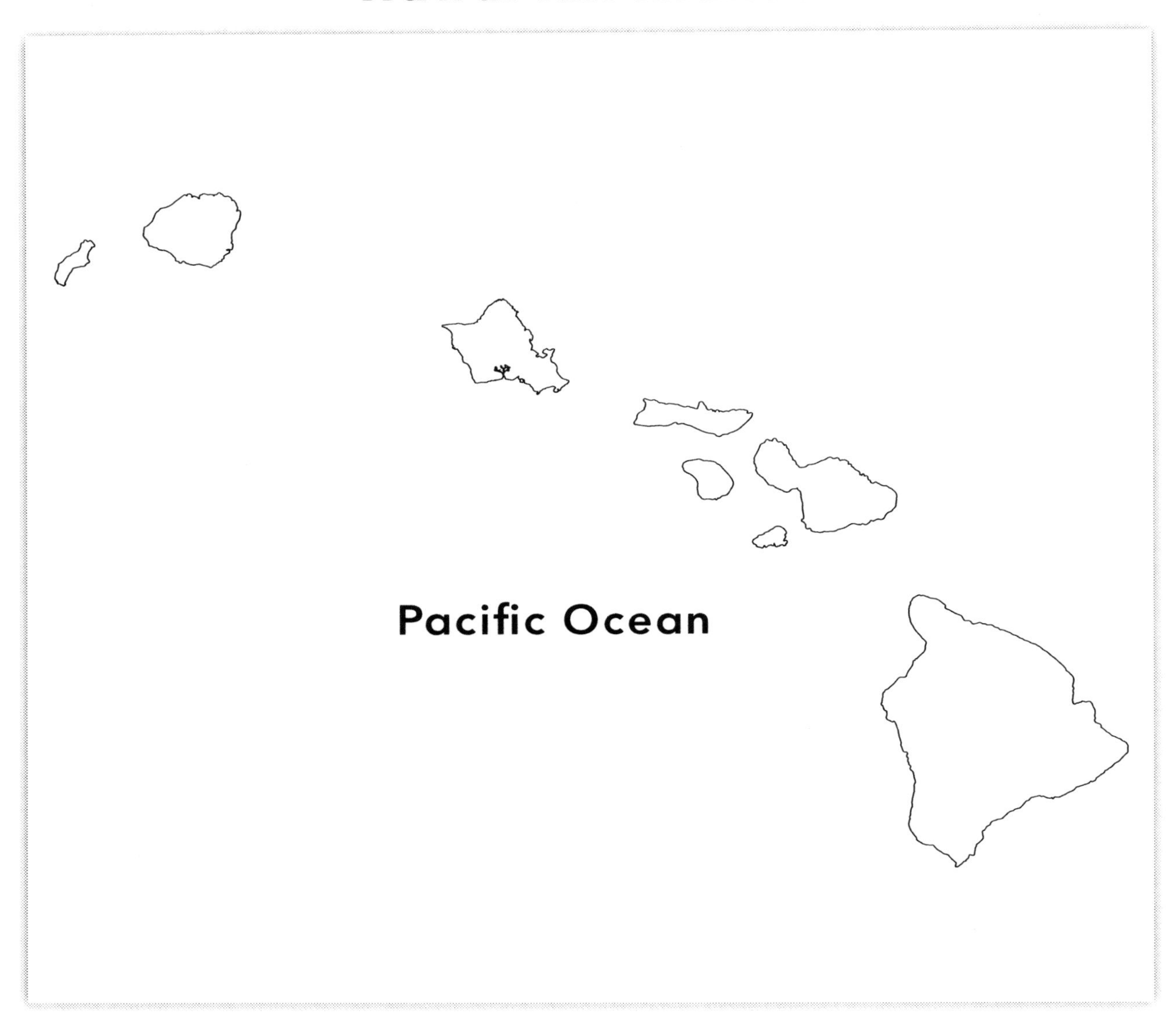

1. Label the island chain. Use the map title to help you.
2. Color the land green.
3. Color the ocean blue.

Name: ______________________ **Date:** ____________

Directions: Read the text. Study the photo. Then, answer the questions.

An Island Chain

The Hawai'ian islands are in the middle of the ocean. They are far from other land. They formed at a hot spot. That is a place below Earth's surface. There, magma rises up through Earth's crust. It oozes into the ocean. Then, it builds up in the water. It makes a volcanic mountain. As it cools, it gets hard like rock. That rock is now land. Only a small part of it sticks out of the ocean. That part is an island.

1. What is a hot spot?

____

____

2. Describe an island.

____

____

Think About It

Name: ______________________________ **Date:** ____________

Directions: This picture shows a hot spot. Study the picture. Then, follow the steps.

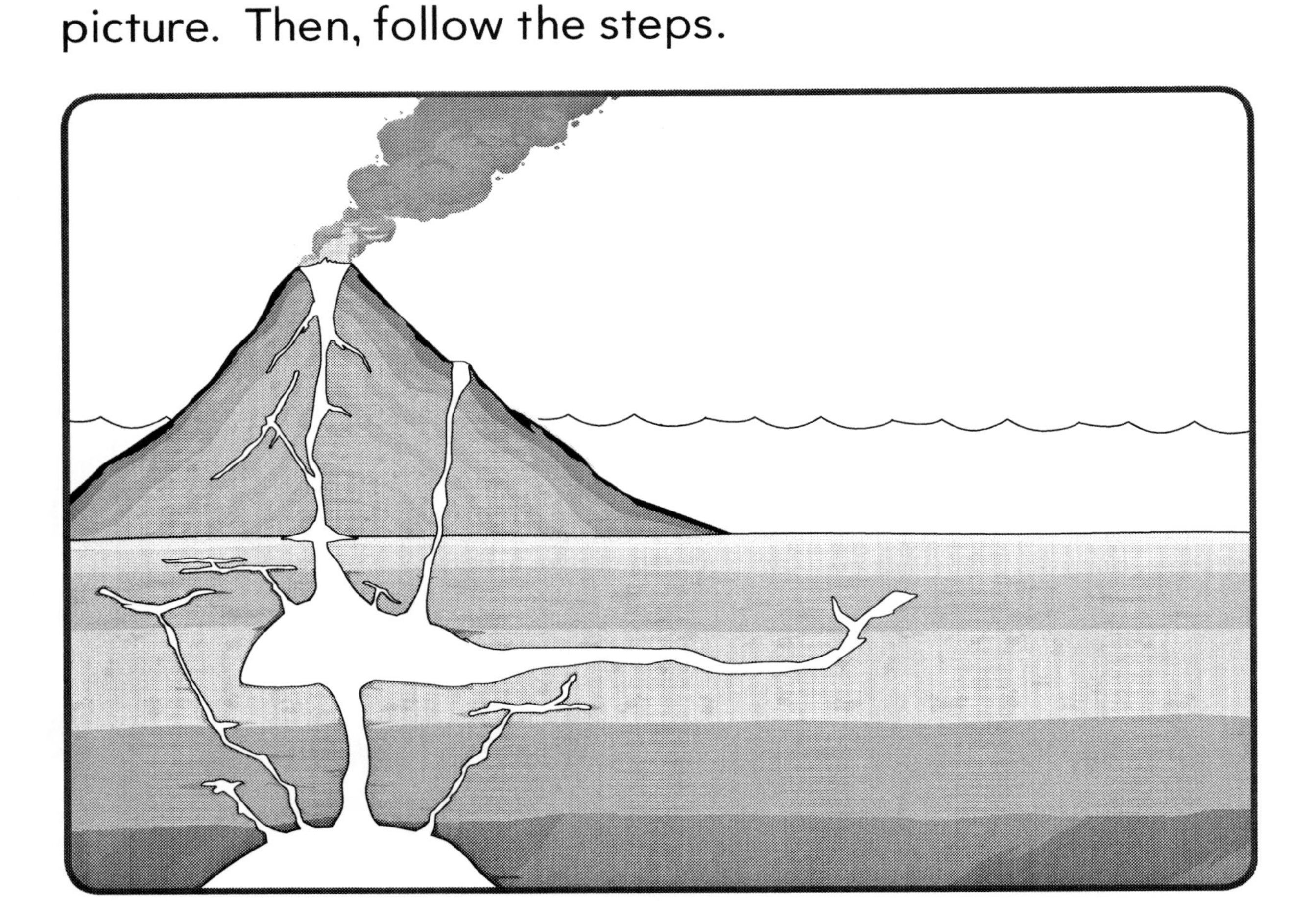

1. The magma is oozing out of the earth's crust. Color it orange.
2. Draw where another island might start from.
3. Part of the island is below water. Part is above water. Label the parts that are below and above the water.
4. Tell a friend how the island grew.

Name: ______________________ **Date:** ______________

Directions: What would a Hawai'ian island look like if you were in an airplane? Draw and write your answer.

Name: ______________________ Date: ____________

Reading Maps

Directions: This is a map of a school. Study the map. Then, answer the questions.

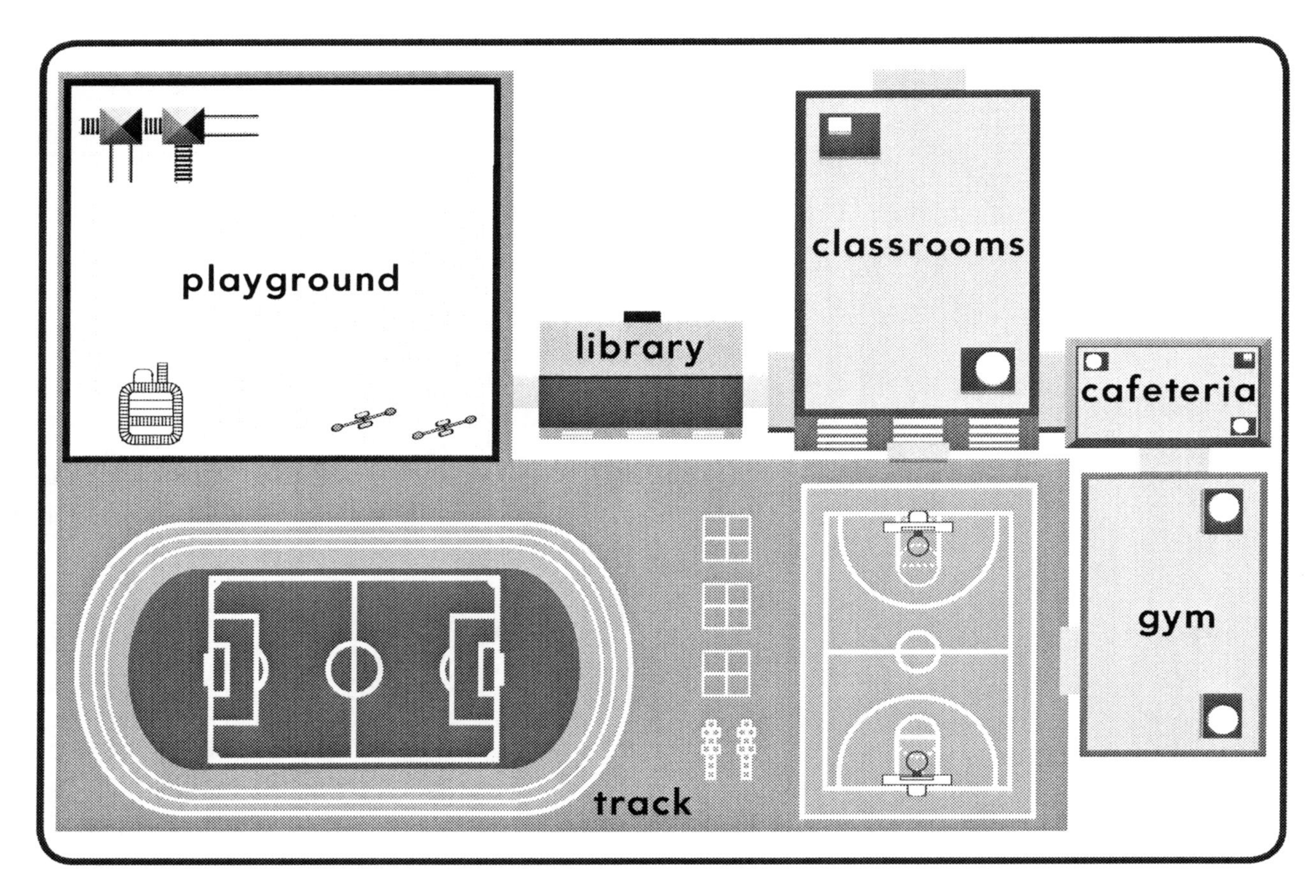

1. Name the places where kids might play games at school.

____

____

2. Circle the classrooms.

3. Draw kids playing a game on the playground.

Name: ______________________ **Date:** ____________

Directions: It is recess time! Draw a route from the classrooms out to the playground. Write directions telling how the kids can get to the playground.

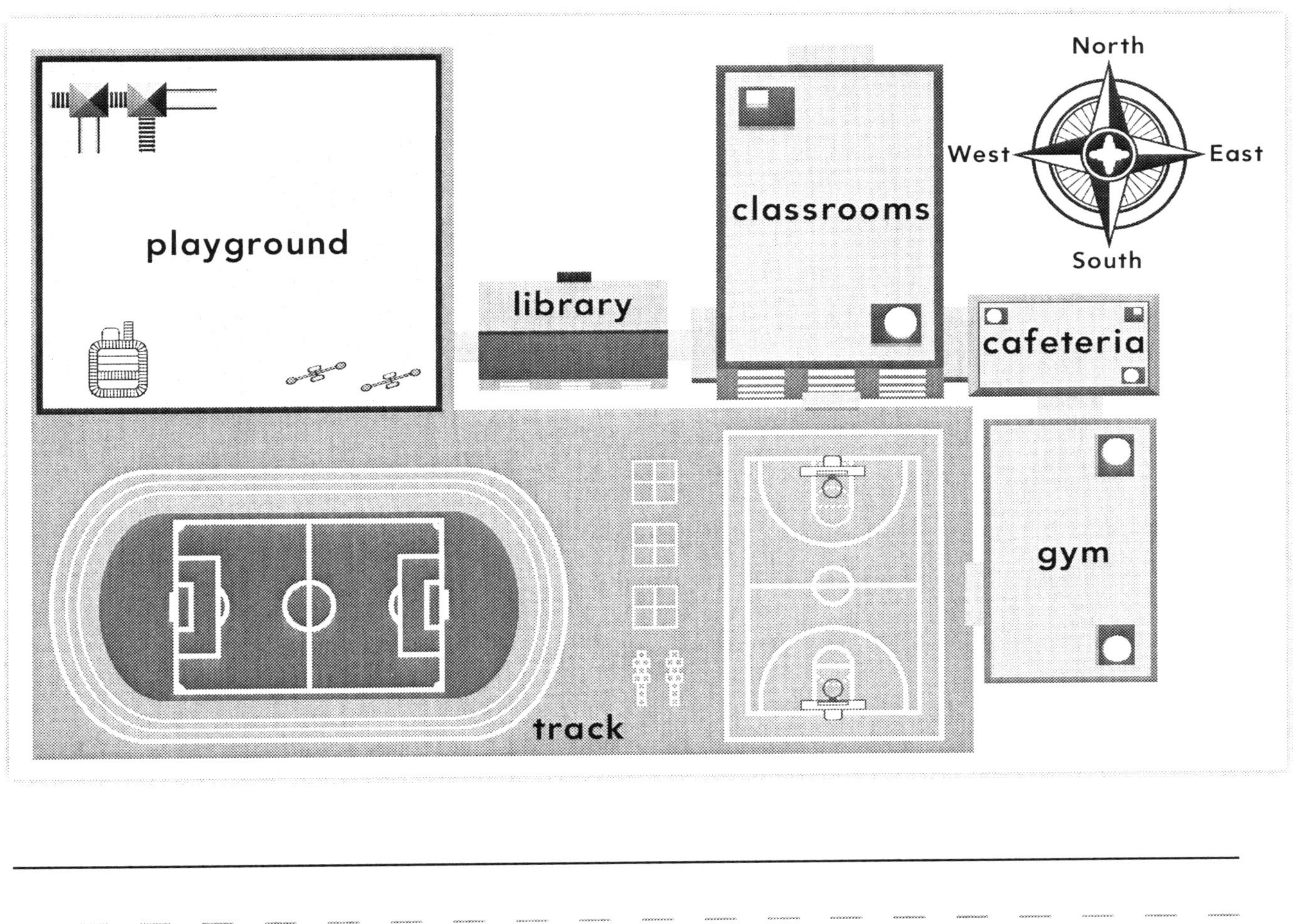

____

____

____

____

Read About It

Name: ______________________ **Date:** ____________

Directions: Read the text. Study the photo. Then, answer the questions.

Games Around the World

Certain games are enjoyed all over the world. Kids play tug of war and hand clapping games. In China, you can play Catch the Dragon's Tail. To play, make a line with other kids. Next, put your hands on the shoulders in front of you. The line leader is the dragon's head. And the last kid in line is the tail. The head tries to catch the tail. All the other kids hold on tight!

1. What kinds of games do kids like to play all over the world?

____

2. Is Catch the Dragon's Tail a game you could play with your whole class? Why or why not?

____

____

Name: ______________________ Date: ____________

Directions: A class votes to decide what game to play at recess. Study the graph. Then, answer the questions.

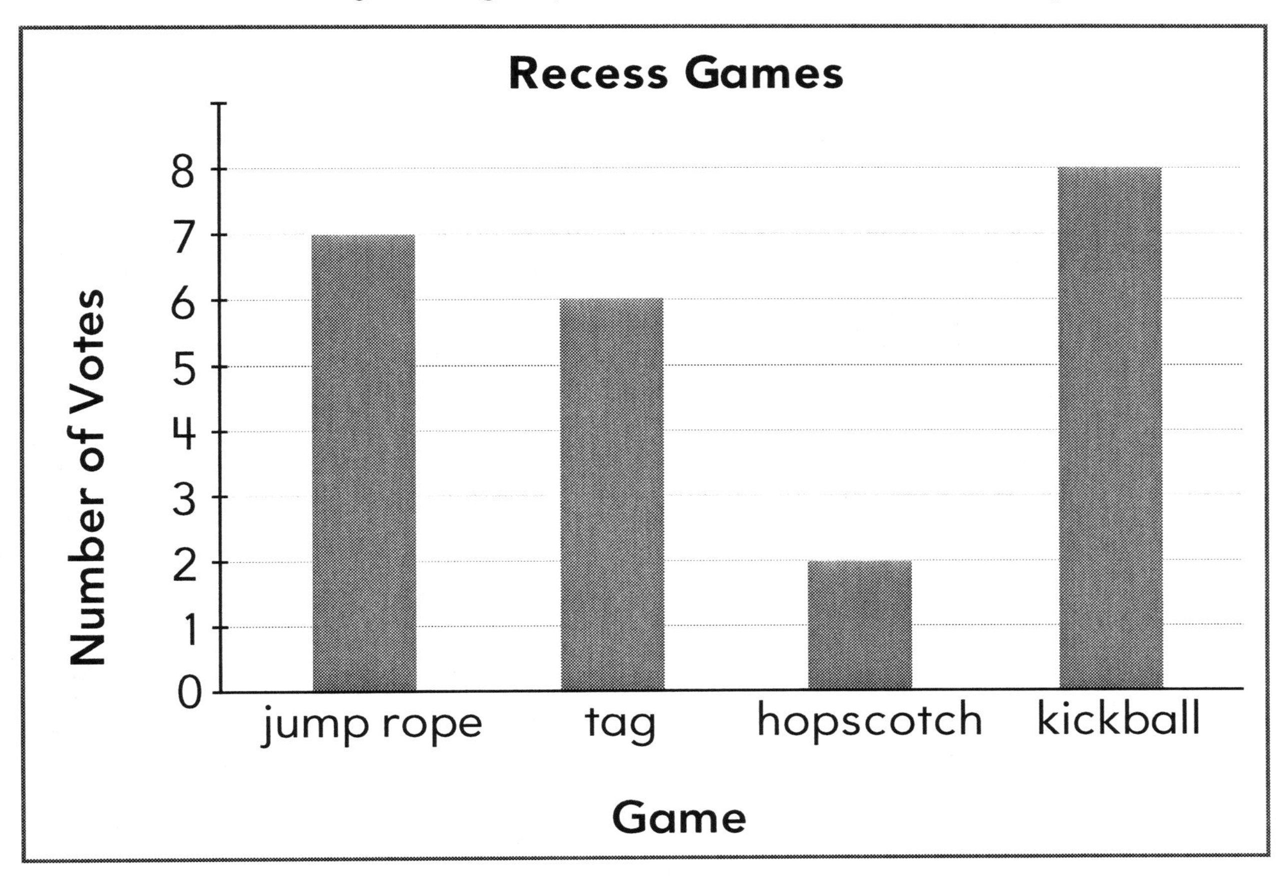

Think About It

1. How many fewer votes did hopscotch get than kickball?

____

2. Why do you think kickball got the most votes?

____

____

Name: ______________________ **Date:** ____________

Directions: Draw and write about your favorite game to play with friends.

Reading Maps

Name: ______________________ **Date:** ____________

Directions: There is a toy hidden somewhere on the playground. Read the clues, and answer the questions.

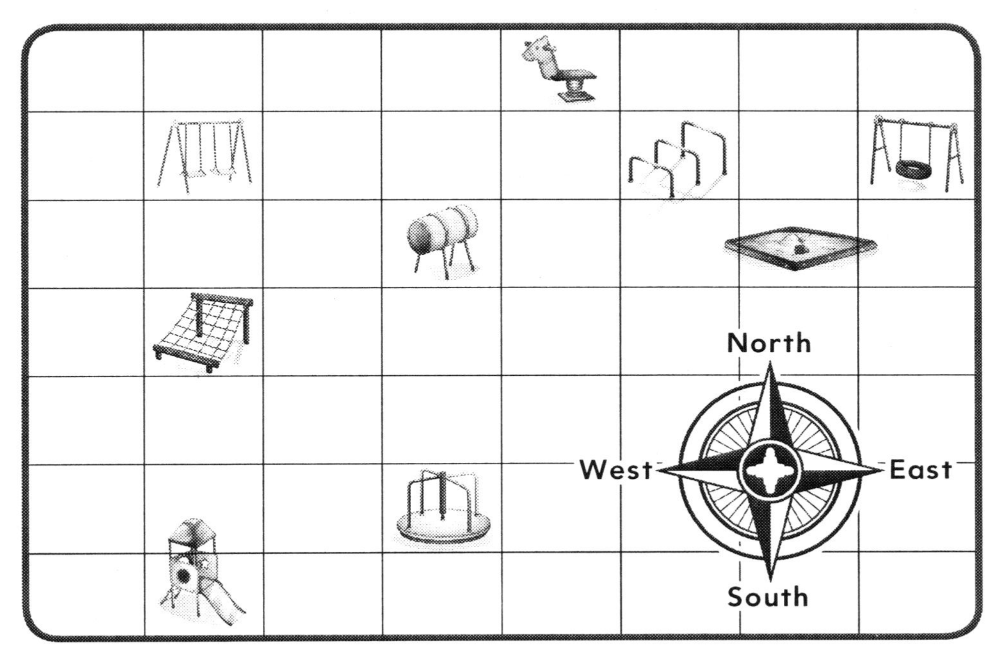

1. Start at the slide. Walk one square north and two squares east. Where are you?

____

2. Walk three squares north and then one square west. What are three things closest to you?

____

3. Walk one square west and one square north. The toy is hidden here. Where are you?

____

Creating Maps

Name: ______________________ **Date:** ____________

Directions: It is your turn to hide a toy on the playground. Pick a spot to hide it. Write clues that a classmate can use to find the hidden toy.

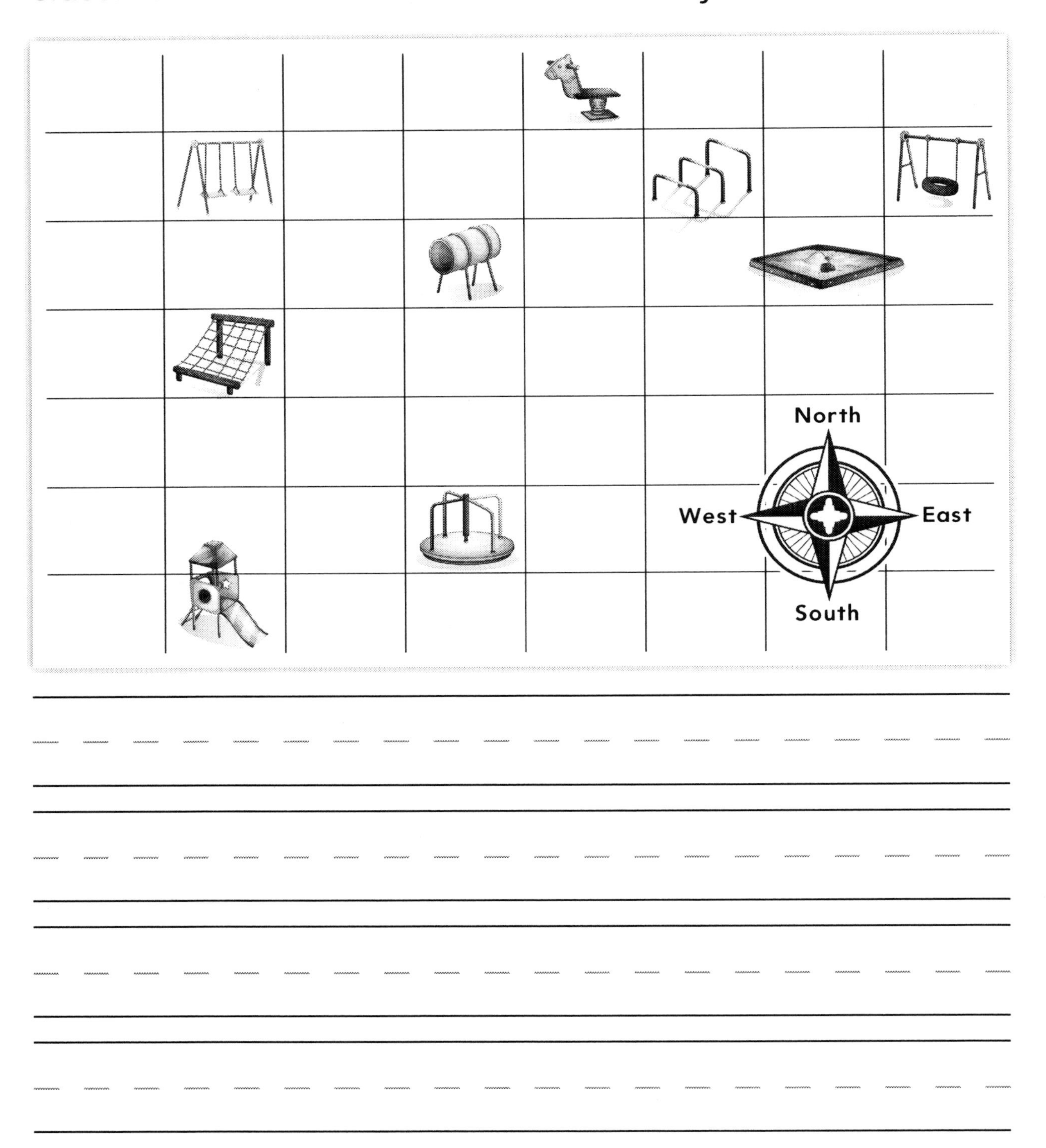

____

____

____

____

Name: ______________________ **Date:** ______________

Directions: Read the text. Study the photo. Then, answer the questions.

Geocaching

Geocaching is like a treasure hunt. People put small things, such as a coin or a charm, in a box. Then, they hide the box outside. What they hide is called a *geocache*. *Geo* means "earth," and *cache* means "a place to keep or hide something." People have made a game out of it. They use GPS devices to keep track of where they hide them. Then, they give people clues online to find them.

1. What is a geocache?

____

____

2. How can geocaching help people get to know their community?

____

____

Think About It

Name: ______________________________ **Date:** ______________

Directions: There are three geocaches hidden in this photo. Use cardinal directions to describe where they are. Then, answer the questions.

1. There is a geocache hidden under a tree. It is in the ______________________________ part of the photo.

2. A geocache is hidden in some bushes. It is in the ______________________________ part of the photo.

3. Someone climbed high to hide this geocache. It is in the ______________________________ part of the photo.

Name: ______________________ Date: ____________

Directions: Draw and write about the best place to hide a geocache in your community.

Reading Maps

Name: ______________________ Date: ____________

Directions: This is a photo of a mountain next to a plain. Study the photo. Then, answer the questions.

1. Describe the shape of the mountain.

____

____

2. How is a mountain different from a plain?

____

____

____

____

3. Create a symbol for a mountain. Create a symbol for a plain.

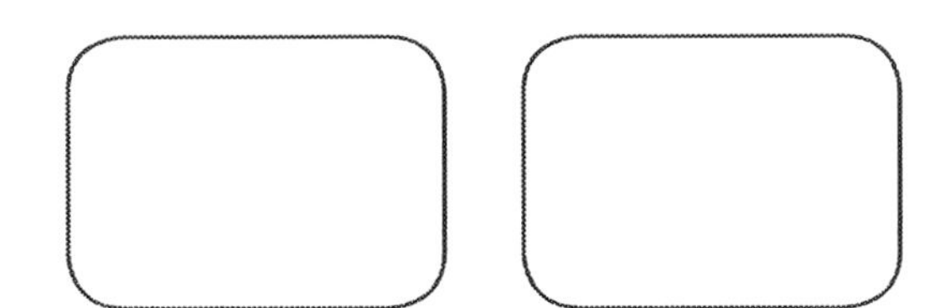

Name: ______________________________ **Date:** ______________

Directions: Draw a map of this location. Include a mountain and plain in your map. Create a map key. Then, give your map a title.

____

Key

Read About It

Name: ______________________ **Date:** ____________

Directions: Read the text. Study the photo. Then, answer the questions.

The Highest Mountain

Mount Everest is the highest mountain in the world. A mountain is a tall and rocky part of land. It is much higher than the land around it. People from all parts of the world want to hike Mount Everest. A mountain guide helps them. But it is hard even with a guide. The air is hard to breathe so high up. It can make people sick. Some people need to turn back. If they do not, they risk their lives!

1. What is a mountain?

____

____

2. Why do you think hikers need a guide to climb Mount Everest?

____

____

Name: ______________________ **Date:** ______________

Directions: This table lists the number of hikers who reached the tops of three mountains. Study the table. Then, answer the questions.

Mountain	Sandia Mountain	Mount Tanen	Mather Mountain
Elevation	11,000 feet	6,680 feet	7,650 feet
Number of Hikers	350	2,100	750

1. Which mountain do you think is the hardest to hike?

____

2. Which mountain do you think is the easiest to hike? How do you know?

____

____

3. How many people get to the top of Mather Mountain each year?

____

Name: ______________________________ **Date:** ______________

Directions: Think of somewhere you could hike in your community. Draw and write about how that place compares to Mount Everest.

Name: ______________________ **Date:** ____________

Directions: This map shows two trails up a mountain. Study the map. Then, answer the questions.

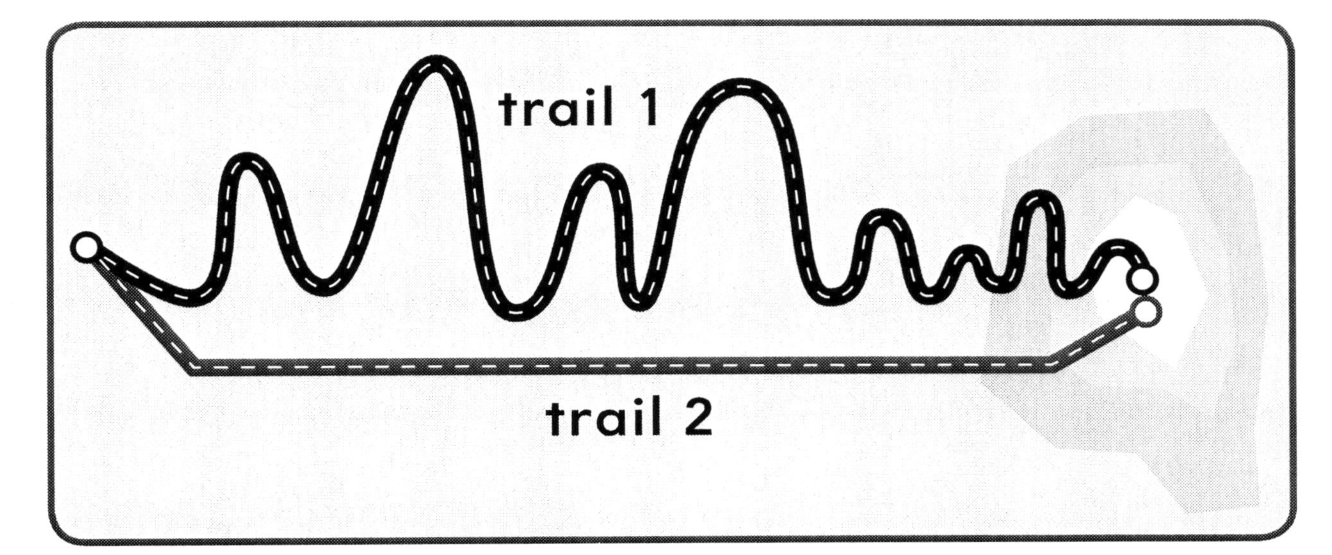

1. Which trail is the shortest route to the peak? How do you know?

2. Which trail do you think would take longer to hike? Explain your answer.

Name: ______________________________ **Date:** ______________

Directions: This map shows two trails up a mountain. Follow the steps to complete the map.

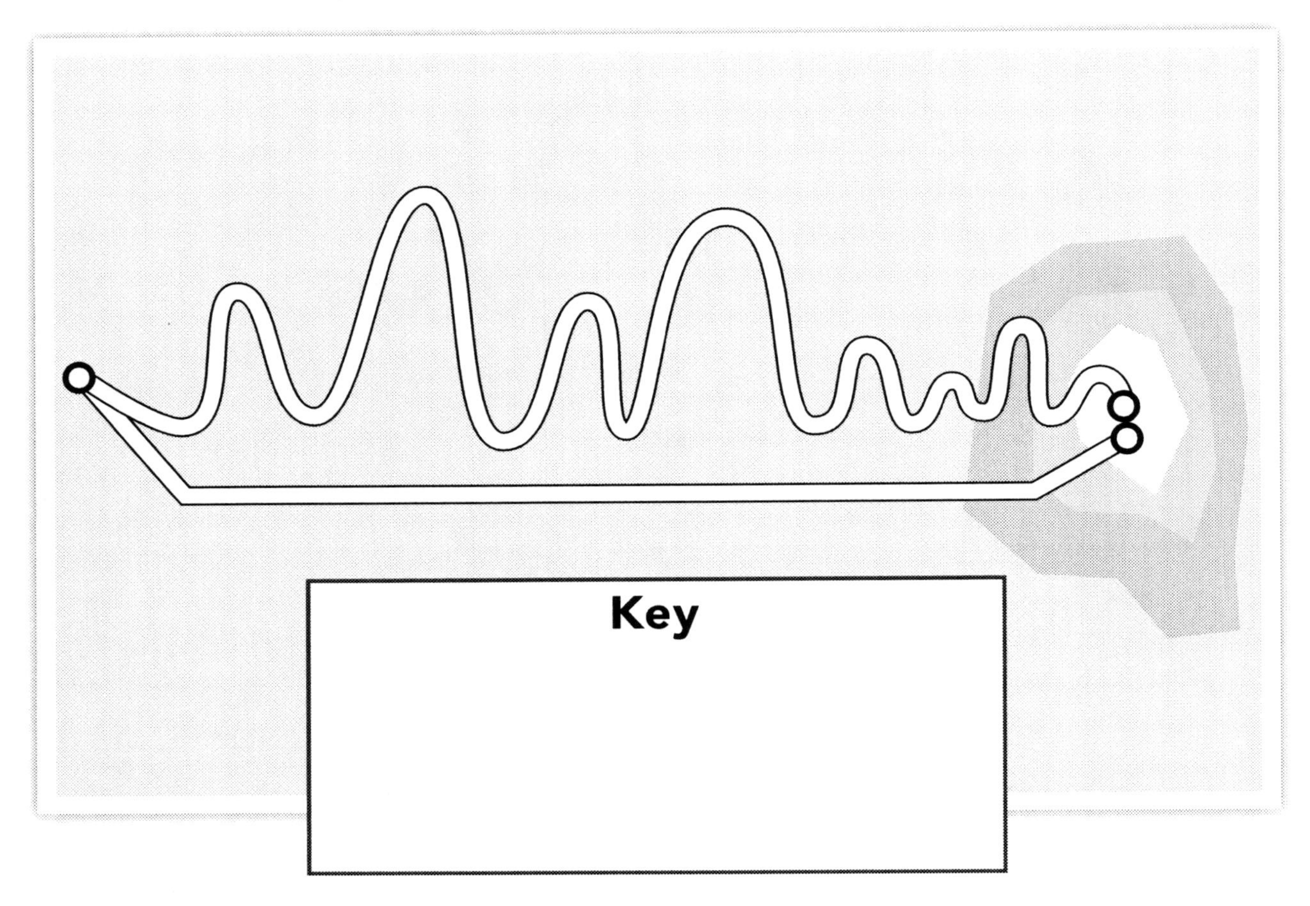

1. Draw a star at the start of the trails.
2. Circle where the trails end.
3. Trace each trail in a different color.
4. Create a name for each trail. Label the trails with the names you created.
5. Create a map key with your colors and trail names.

Name: ______________________ **Date:** ____________

Directions: Read the text. Study the photo. Then, answer the questions.

A Short Route

Sometimes, a short route is not a fast one. The Incline is a trail for hikers. It is in Colorado. And it is a steep trail! It goes straight up the side of the mountain. But it is less than a mile to the top. Most people can walk a mile in 15 to 20 minutes. But on the Incline, most hikers need an hour or more to reach the top. It is so steep that they need to go slow.

1. Why does it take people so long to hike the Incline?

____

____

2. Is the shortest route always the fastest route? How do you know?

____

____

Think About It

Name: ______________________ **Date:** ____________

Directions: Each trail leads from the bottom to the top of Animal Mountain. Study the table. Then, answer the questions.

Trail	Distance to the Top
Rabbit Trail	1 mile
Deer Trail	5 miles
Moose Trail	3 miles

1. Which trail would be the shortest route to the top?

____

2. Which trail is likely the easiest to hike?

____

3. Draw Animal Mountain with all three trails.

Name: ______________________________ **Date:** ______________

Directions: Would you rather hike a short and steep route or a long route to the top of a mountain. Why? Draw and write your answer.

Reading Maps

Name: ______________________ **Date:** ____________

Directions: This map shows some of the volcanoes in the United States. Study the map. Then, answer the questions.

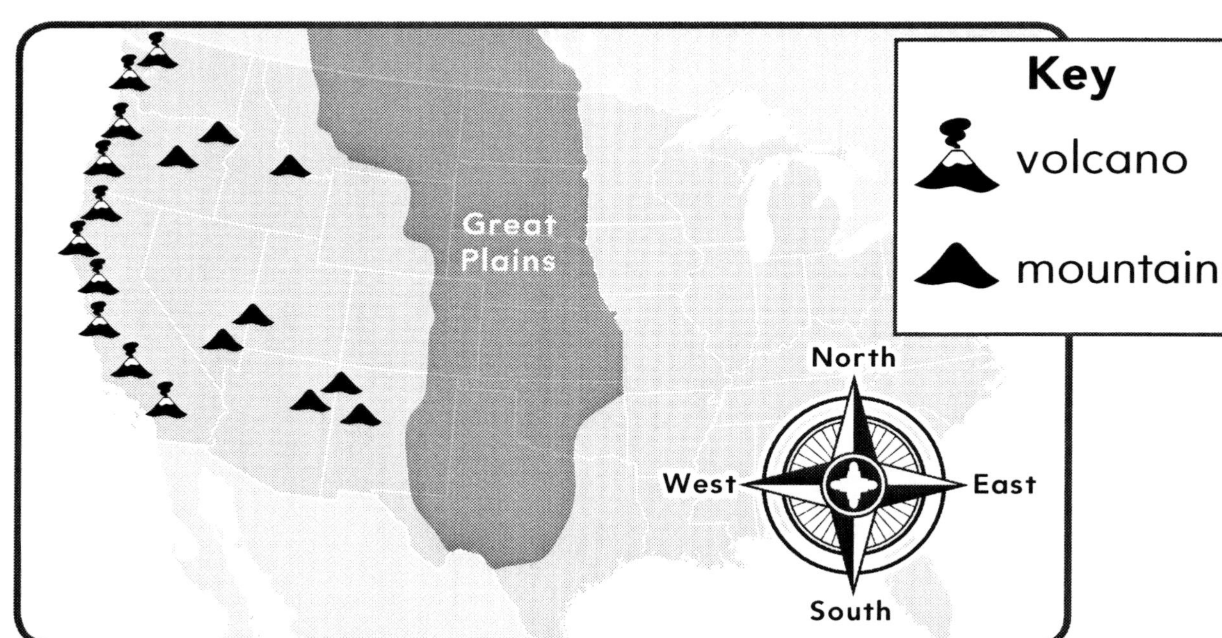

1. The most volcanoes are in the ______________ part of the map.

2. What part of the United States has no volcanoes?

____

3. Are there volcanoes in the Great Plains? How do you know?

____

____

Name: ______________________________ **Date:** ____________

Directions: Imagine there is a volcano erupting in the United States. Use the clues to find out where.

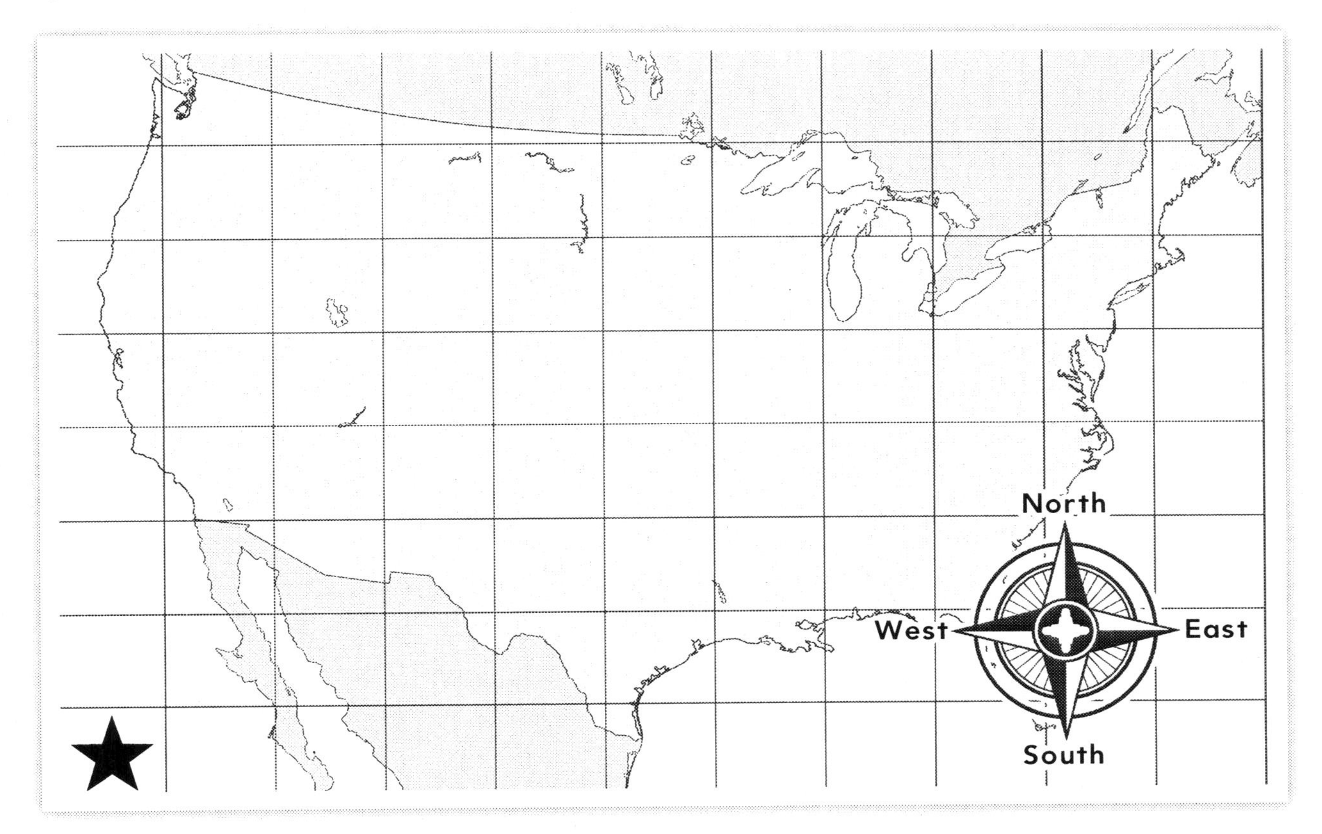

1. Start at the star.
2. Go north six squares.
3. Go east two squares.
4. Draw the symbol of a volcano in that spot. Label it *Active Volcano*.
5. Tell a friend what part of the country the volcano is in.

Name: ______________________ Date: ____________

Directions: Read the text. Study the photo. Then, answer the questions.

Read About It

Mount St. Helens

Mount St. Helens looks like a mountain. But it is a tall volcano. The top of it is missing. How did that happen? First, an earthquake shook the land. That caused a huge chunk of the mountain to break off. It tumbled down the slope at top speed. Then, lava shot out of the vent. In the photo, you can see the rock and lava that poured down the side.

1. How is a volcano similar to a mountain? How is it different?

____

____

2. Explain why the top of the volcano is missing.

____

____

Name: ______________________ Date: ____________

Directions: This is a diagram of a volcano. Study the diagram. Answer the questions, and follow the steps.

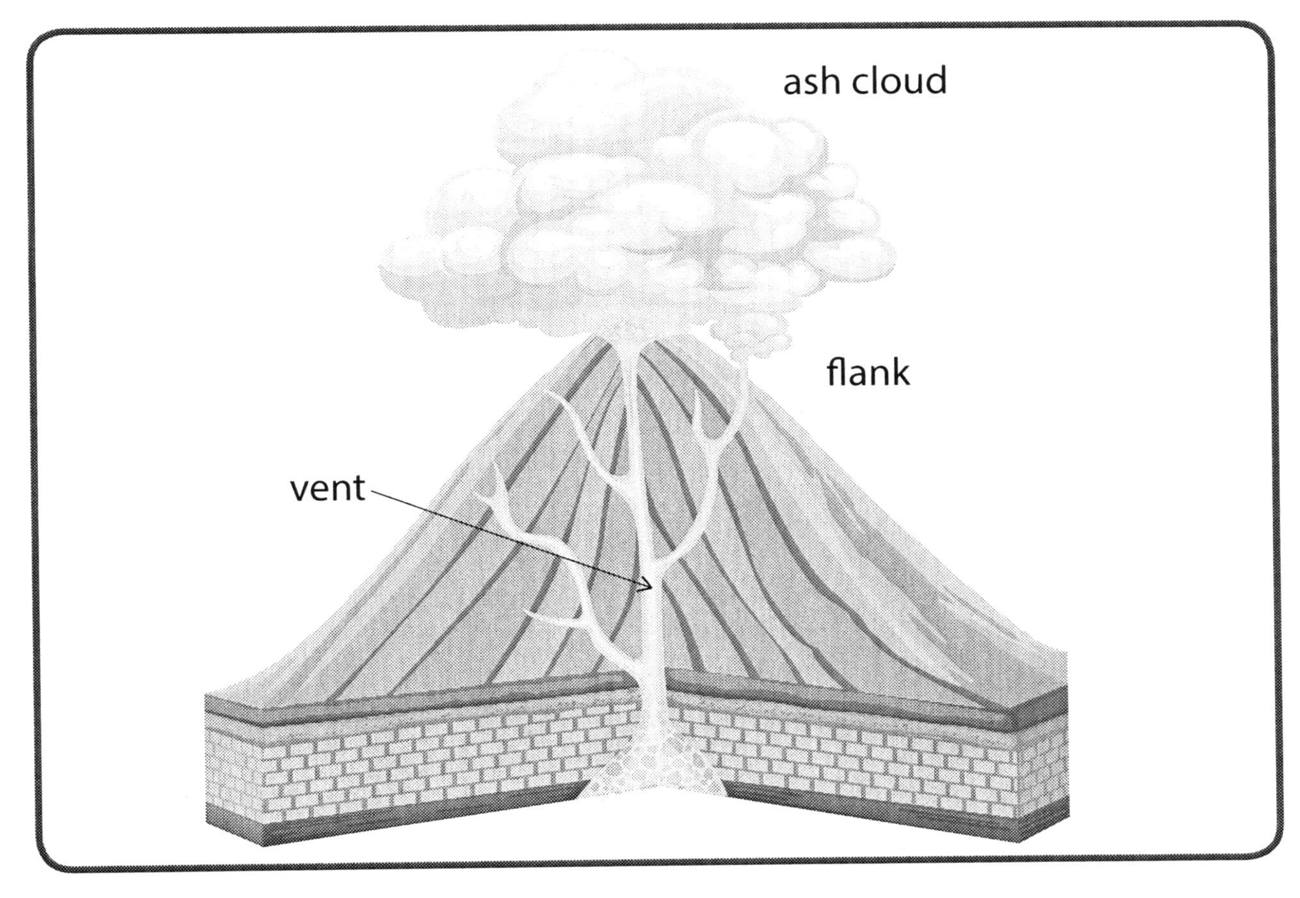

1. Study the shape of the volcano. What other landforms have a similar shape?

____

____

2. Magma is found inside of the volcano. Label the magma.

3. When magma reaches the surface, it is called *lava*. Draw lava on the outside of the volcano.

4. Draw an arrow pointing to where lava erupts.

Geography and Me

Name: ______________________ **Date:** ______________

Directions: Would you feel safe living near a volcano? Why or why not? Draw and write your answer.

Name: ______________________ **Date:** ____________

Directions: This is a cliff. A cliff is a landform. Study the photo. Then, answer the questions.

1. Describe the environment where this cliff formed.

____

____

2. The cliff wall is the vertical side of the cliff. Shade the cliff wall blue.

3. Draw a green arrow pointing to the bottom of the cliff.

4. Draw a red arrow pointing to the top of the cliff.

Creating Maps

Name: ______________________ **Date:** ____________

Directions: Draw a diagram of a cliff based on the photo. Label the top of the cliff, the cliff wall, and the bottom of the cliff.

Name: ______________________ Date: ____________

Directions: Read the text. Study the photo. Then, answer the questions.

Read About It

Lighthouse at the Cliff

A cliff is a landform made of rocks. It has steep walls. It is a big drop from the top to the bottom. A foggy night makes the cliff hard to see. Boats crashed into the side of the rocky cliff without a warning. A lighthouse stands tall. This landmark is there to shine a light out to the ocean. On foggy nights, boat captains can see the light. They can steer their boats away from the shore.

1. What can happen to a boat on a foggy night?

____

____

2. How does a lighthouse help a boat?

____

____

Name: ______________________ **Date:** ____________

Directions: A landslide happens when part of the land gives way. All the dirt and rock fall to the land below. Study the diagram. Then, answer the questions.

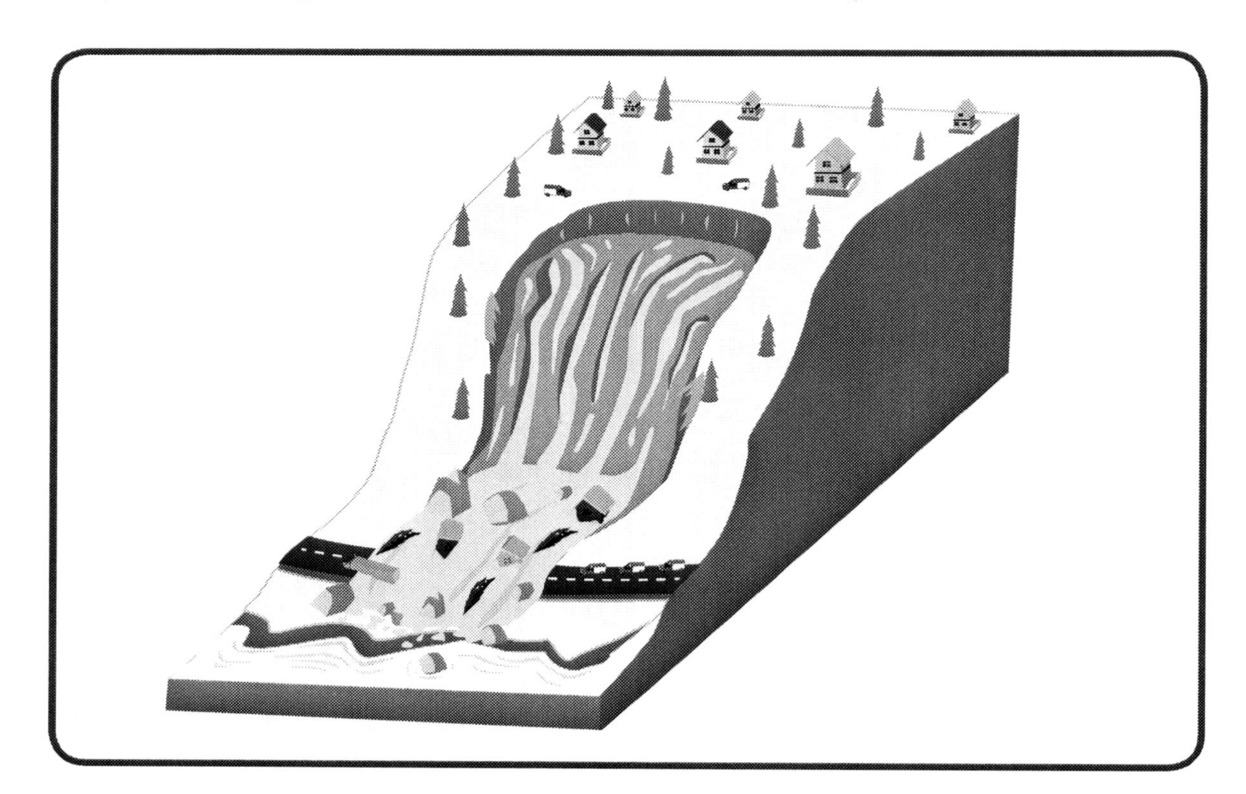

1. Circle the landslide.

2. Do you think houses should be built on the edge of a cliff? Why or why not?

Name: ______________________ **Date:** ______________

Directions: Would you build your house on a cliff? Why or why not? Draw and write your answer.

Name: ______________________ **Date:** ____________

Directions: This map shows the restaurants in a city. Study the map. Then, answer the questions.

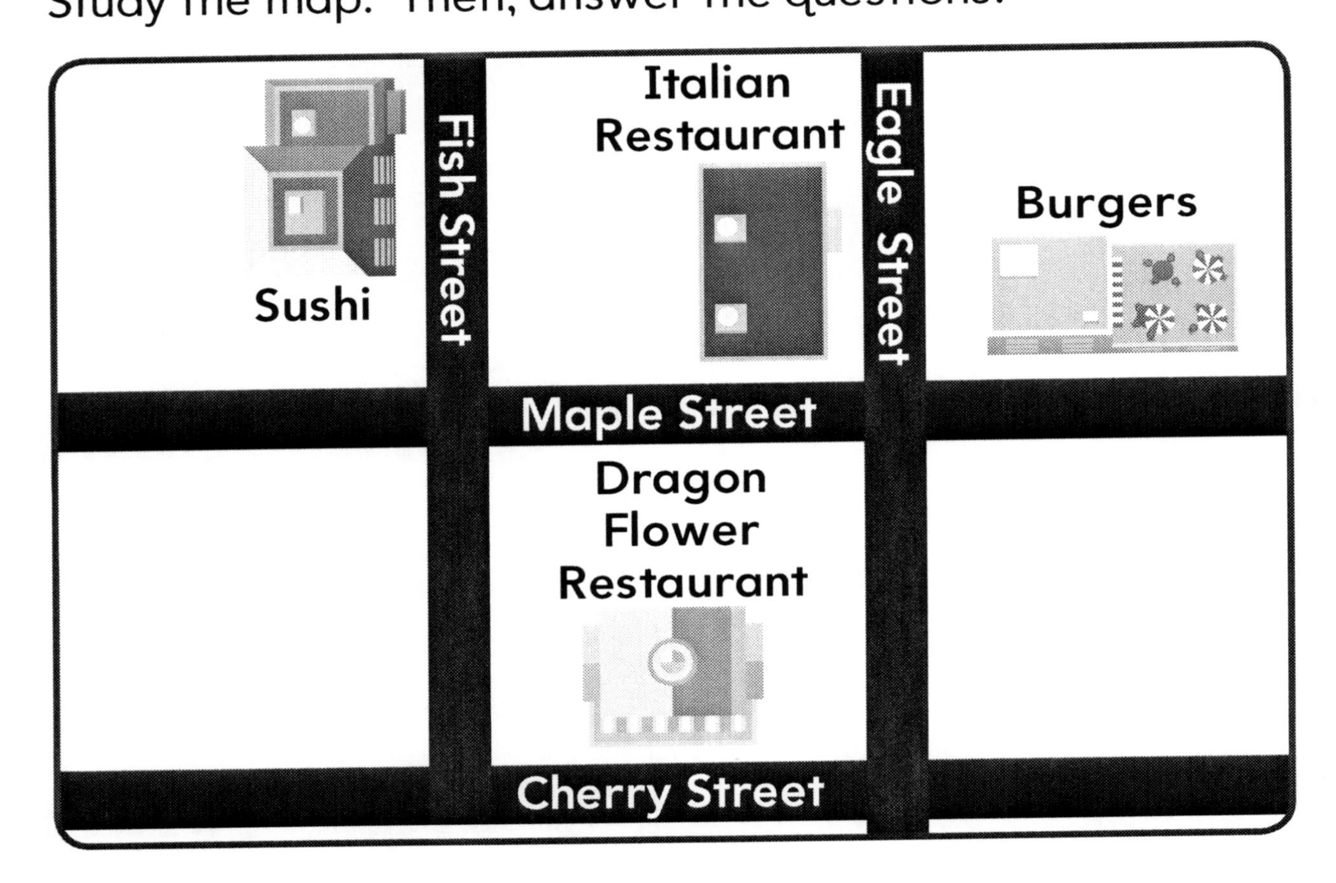

1. What street is the Italian restaurant on?

____

2. There is a Chinese restaurant on Cherry Street. Circle the restaurant. What is the name of the restaurant?

____

3. Tell a friend which restaurant you would like best.

Name: ______________________ Date: ____________

Directions: Joe is walking down Eagle Street. Draw the route he needs to take to get to the sushi restaurant. Then, write directions that he can follow.

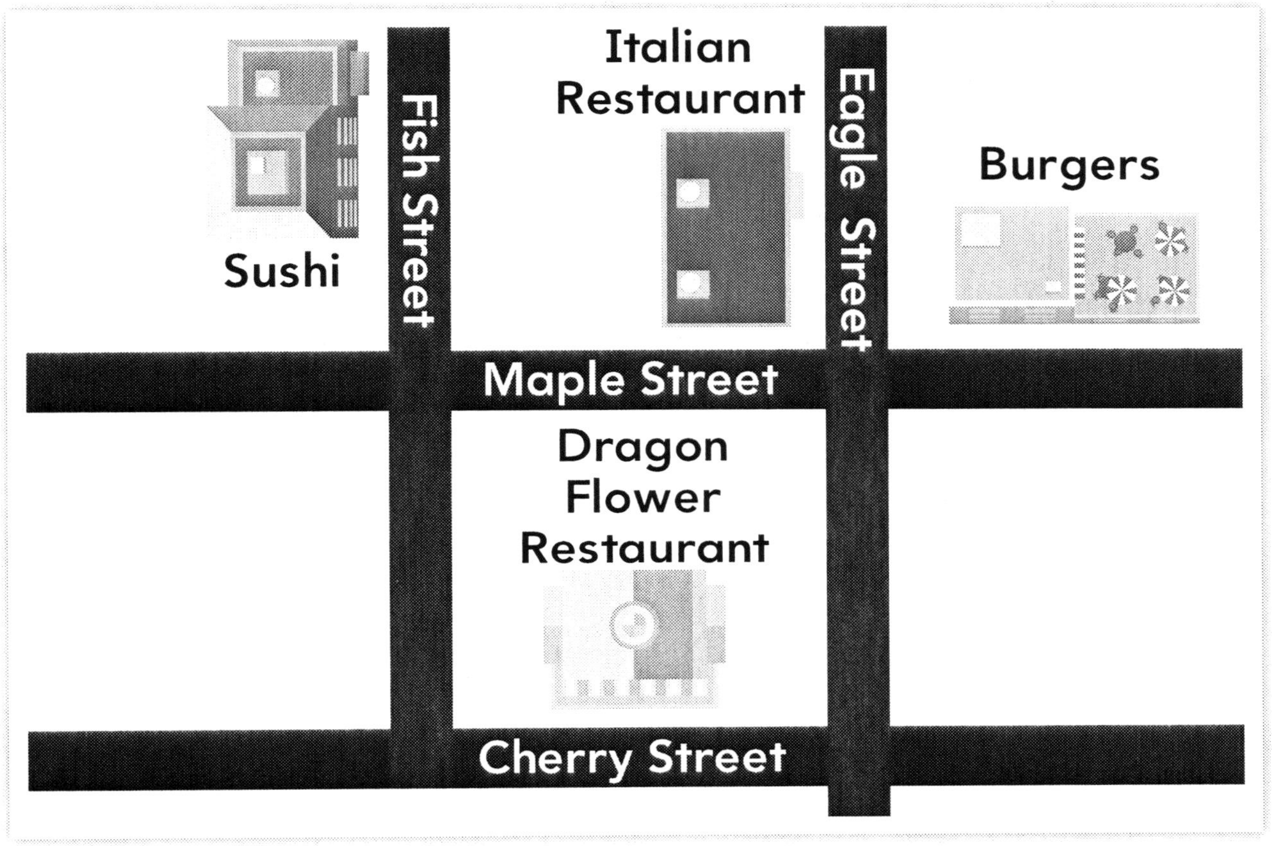

____

____

____

____

Name: ______________________ **Date:** ____________

Directions: Read the text. Study the photo. Then, answer the questions.

Food from Around the World

You can find many kinds of food in the city. People move to the city from other countries. They love to share the food from their home countries. Some people open up restaurants. They make food using their old family recipes.

A street fair is a good place to eat. There are many booths there. The food is from all over the world. Visit all the carts and booths to get a tasty bite.

1. Why are there many kinds of food in the city?

____

____

2. Why is a street fair a good place to eat?

____

____

Name: ______________________ **Date:** ____________

Directions: A city is having a street fair. Each booth will sell food from a different country. Study the map. Then, answer the questions.

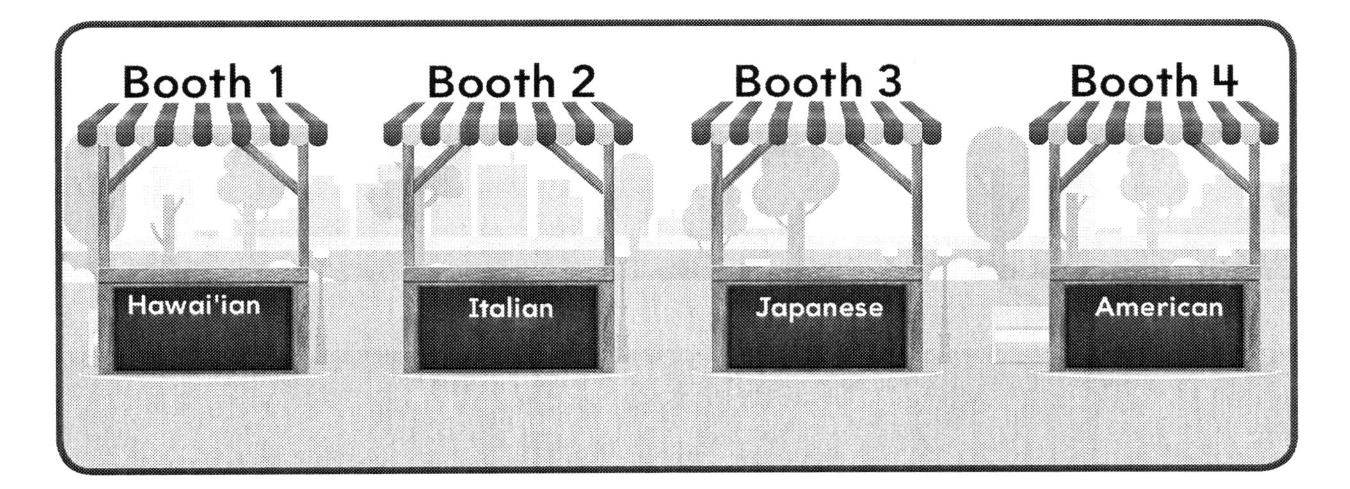

1. What type of food is being sold at Booth 3?

____

2. If you want to buy Italian food, which booth should you visit?

____

3. List the types of food that are at the street fair.

____

____

Name: ______________________________ **Date:** ______________

Directions: Draw and write about a new food you have tried.

ANSWER KEY

There are many open-ended pages and writing prompts in this book. For those activities, the answers will vary. Answers are only given in this answer key if they are specific.

Week 1 Day 1 (page 15)

1. My Bedroom
2. under
3. Students should draw a car on top of the rug.

Week 1 Day 2 (page 16)

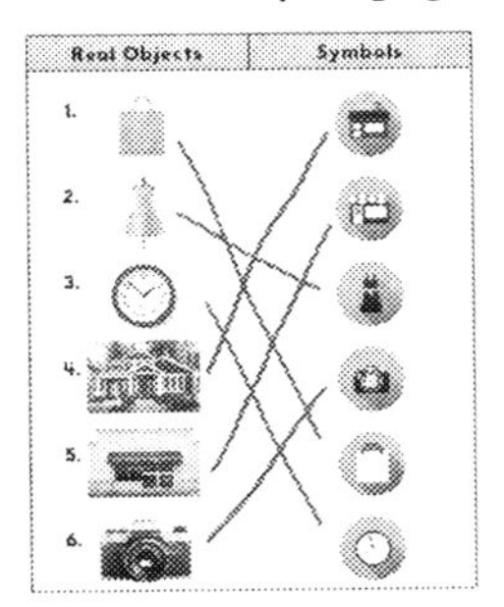

Week 1 Day 3 (page 17)

1. Students should circle the symbol for the park.
2. Students should draw the symbol for the house in three places.
3. Students should draw a box around the symbol for the zoo.
4. Students should create a symbol for an ice cream store and draw it on the map.

Week 1 Day 4 (page 18)

1. stairs or climbing ramp
2. west
3. Example: *Never eat Shredded Wheat*.

Week 1 Day 5 (page 19)

1. Students should have labeled the compass rose.
2. mountain, treasure, octopus
3. east
4. west

Week 2 Day 1 (page 20)

1. east, north, right

Week 2 Day 3 (page 22)

1. Ant Street
2. Main Street
3. First Street

Week 2 Day 5 (page 24)

1. The small building at the top left should be colored.
2. stadium
3. The building at the bottom right should be colored.
4. north

Week 3 Day 3 (page 27)

1. An aerial view is looking at the ground from up high.
2. Houses look like tiny boxes.

Week 4 Day 1 (page 30)

1. beach, park, or pool
2. ice cream shop
3. There are two gas stations.

Week 4 Day 3 (page 32)

1. Answers may include that people can go to stores, parks, and special places like an airport or a museum.
2. Yes, a museum is a part of a community because it is a special place people can go.

Week 4 Day 4 (page 33)

1. People can go to a farmers' market to buy things, such as fruit and vegetables.
2. Answers may include lettuce, cucumbers, bell peppers, or peas.

Week 5 Day 1 (page 35)

1. The park is farther from Maria's house.
2. It would take Maria longer to walk to the store because it is farther from her house.

Week 5 Day 3 (page 37)

1. People move to a new country because it will give them a better life.
2. When people move, they have to find a new house and job. They will make new friends.

Week 5 Day 4 (page 38)

1. Marcus
2. Canada
3. Answers may include that he is nervous, excited, or misses home.

ANSWER KEY *(cont.)*

Week 6 Day 1 (page 40)

1. Park Street
2. Answers may include any of the following: all the numbers start small and get bigger; on one side of the street the numbers are odd; on the other side of the street they are even; they go up one number for each house.

Week 6 Day 3 (page 42)

1. An address tells the street and house number. It also tells the city and state.
2. Every place has its own address to make it easy to find.

Week 6 Day 4 (page 43)

1. Penny Clemons
2. No, the letter could not be delivered. The mail carrier would not know which house on Eagle Street to deliver it to.

Week 7 Day 1 (page 45)

1. Example: *Walk down Main St. Turn right on Pine St.*
2. Example: *Walk down Main St. Pass Pine St.*

Week 7 Day 3 (page 47)

1. A route is the way you go to get somewhere.
2. Answers may include that people use GPS to track where they walk, drivers use it for directions, and people track pets.

Week 7 Day 4 (page 48)

1. Fur St.
2. Maple St., Ash St., Tail St., Redwood St., and Fur St.
3. Students should draw a line down Fur Street and straight east on Maple Street.

Week 8 Day 1 (page 50)

1. The North Pole is at the top of the globe.
2. The South Pole is at the bottom of the globe.
3. Students should use a red crayon to trace the dotted line on the center of the globe.

Week 8 Day 3 (page 52)

1. Places close to the equator get more sunlight.
2. equator

Week 8 Day 4 (page 53)

1. North and South Poles: snowman, jacket, hat
   Equator: sunscreen, shorts, flip flops
2. Answers may include swimming suits, sandals, and sun hats.

Week 9 Day 1 (page 55)

1. Students should have drawn a line on the photo dividing the rural and urban communities.
2. One community has a lot of houses in it. The other has a lot of land.

Week 9 Day 2 (page 56)

The farm should be labeled on the left, and the road should be labeled on the right.

Week 9 Day 3 (page 57)

1. The land is flat. There is wheat growing there.
2. Answers may include bread, baked goods, cereal, and pasta.

Week 9 Day 4 (page 58)

Rural: cow, tractor, and barn
Not Rural: apartment building and taxi

1. Those are things found on a farm. Farms are in rural communities.
2. Answers may include dirt roads, farm animals, crops, and farmland.

Week 10 Day 1 (page 60)

1. Answers may include a mountain, a river, a lake, and a forest.
2. The area with many trees should be circled.
3. Symbols should represent each item listed.

Week 10 Day 2 (page 61)

1. mountain
2. lake
3. forest
4. river

Week 10 Day 3 (page 62)

1. A plateau is tall and flat on top. The plains are flat and do not rise up.
2. A plateau is like a mountain because it is tall. A plateau is flat on the top.

ANSWER KEY *(cont.)*

Week 10 Day 4 (page 63)

1. Answers may include a plain that looks like flat land or a hill that looks like a smaller mountain.
2. The river leads to the lake.
3. A hill is not as tall as a mountain.

Week 11 Day 1 (page 65)

1. It shows fish, timber, deer, water, and crops.
2. People can get water and fish from the river.

Week 11 Day 3 (page 67)

1. A truck takes them to the sawmill.
2. Trees do not run out. More can be planted.

Week 11 Day 4 (page 68)

wood products: cutting board, pencil, spoon, table
not wood products: toaster, trash can, pan, oven mitts

1. Students should list which wood products they have at home.
2. Spoons, cutting boards, and tables can be made from other resoruces.

Week 12 Day 1 (page 70)

1. The size of the ocean is wider and larger than the lake.
2. The shape of the river is long and skinny. The lake is much wider and not as long.

Week 12 Day 2 (page 71)

From top to bottom: lake, river, ocean

Week 12 Day 3 (page 72)

1. Rivers bring water into the lake.
2. There are islands in the middle of Lake Baikal.

Week 12 Day 4 (page 73)

1. Students should trace the river in the top-left corner of the map.
2. Answers may include that land is shown on all sides of the lake or that all the edges of the lake can be seen.

Week 13 Day 1 (page 75)

1. The canyon looks deep and wide.
2. A river is at the bottom of the Grand Canyon.

Week 13 Day 3 (page 77)

1. A river is powerful because it can cut through rock.
2. A canyon forms from a river cutting through the rock.

Week 13 Day 4 (page 78)

1. The canyon walls of China's Grand Canyon are not as steep.
2. The canyons were both made by a river cutting the land. China's Grand Canyon has a lot of trees. The Grand Canyon in the United States does not.

Week 14 Day 1 (page 80)

1. The wind turbine symbol from the map should be drawn.
2. mountain and lake
3. They are in the center of the map.

Week 14 Day 2 (page 81)

A wind turbine should be drawn in the upper-left corner.

Week 14 Day 3 (page 82)

1. Wind farms are built in places with strong and steady wind.
2. Wind is a renewable resource because it will never get used up.

Week 14 Day 4 (page 83)

1. meadow
2. It should not be built on the high hill because it does not have high wind speeds.

Week 15 Day 1 (page 85)

1. Students should have drawn an antenna on a skyscraper and a star in the park.
2. Hudson Street
3. First Street

Week 15 Day 3 (page 87)

1. Shanghai is a crowded place. A lot of people live and work there. It has big buildings.
2. Smog can make it hard to breathe.

Week 15 Day 4 (page 88)

1. They travel in cars, on a bus, on bikes, and on foot.
2. Many people can ride on a bus. It gets cars off the road.

ANSWER KEY *(cont.)*

Week 16 Day 1 (page 90)

1. Students should draw a curved line.
2. Students should circle the mountain symbol.
3. The mountain symbol is taller than the hill symbol. The hill symbol is more rounded.

Week 16 Day 3 (page 92)

1. It can run out.
2. Gold is found in the mountains and river.

Week 16 Day 4 (page 93)

1. jewelry
2. electronics, coins, and business
3. Students should justify their answers with a reason.

Week 17 Day 1 (page 95)

1. They are islands because they are surrounded by water.
2. They could take a boat or walk across a bridge.

Week 17 Day 3 (page 97)

1. People ride boats, walk, and ride bikes in Venice.
2. People need bridges to cross to the water.

Week 18 Day 1 (page 100)

1. Students should describe the mountain, tunnel, or lake.
2. Students should circle the building near the top of the map. Explanations may include that it rises high above the other buildings.

Week 18 Day 3 (page 102)

1. A landmark is something that stands out from a distance.
2. Big Ben is a big clock tower. It is bigger than the other buildings around it.

Week 18 Day 4 (page 103)

1. Students should describe something that stands out in the photo, such as the Ferris wheel.
2. Yes, because it is large and it stands out.

Week 19 Day 2 (page 106)

1. 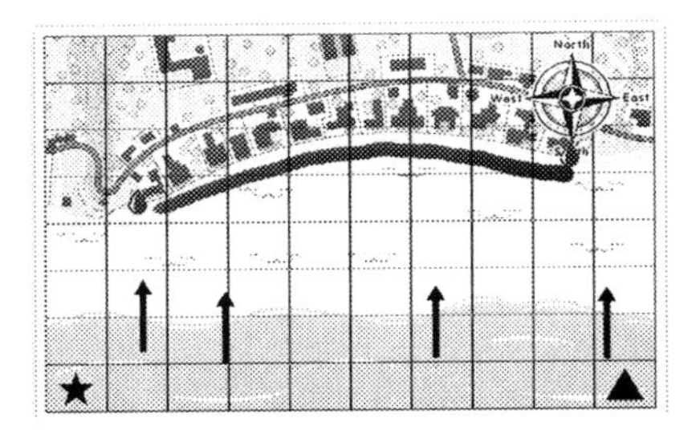

Week 19 Day 3 (page 107)

1. in the ground
2. The environment is covered in oil. The oil gets all over the animals.

Week 19 Day 4 (page 108)

1. Answers may include that oil is messy and it is not good for you to touch.
2. Answers may include that it is hard to clean up because oil is sticky. It is covering the whole beach.

Week 20 Day 1 (page 110)

1. Students should have drawn a circle around the city.
2. Students should have drawn a box around the desert.
3. Answers may include that it is hot or there are fewer resources.

Week 20 Day 2 (page 111)

apartment building: a lot of people
house: a few people
desert: no people

Week 20 Day 3 (page 112)

1. There is a desert in the middle of Australia. There is a rain forest on the coast.
2. It was hard to live there because there were floods, pests, and diseases. People were also far from other towns.

Week 20 Day 4 (page 113)

1. the coast
2. Little to no people live in the center. I looked at the map key.
3. More people live on the east side of Australia.

ANSWER KEY *(cont.)*

Week 21 Day 1 (page 115)

1. Green Stop
2. Line 2 has more stops. It has five stops.

Week 21 Day 2 (page 116)

1. Line 2 crosses the river.
2. Students should have drawn a box on the Blue Stop.
3. Students should have drawn a triangle on the Yellow Stop.
4. Students should have drawn a star on the Green Stop.
5. Red Stop to Green Stop, then Black Stop.

Week 21 Day 3 (page 117)

1. A subway is faster.
2. A subway goes underground.

Week 21 Day 4 (page 118)

1. 6:00 am, 11:00 am
2. 8:00 am has the tallest bar on the graph.
2. Artists come to show and sell their crafts.

Week 22 Day 1 (page 120)

1. Students should color the river blue.
2. Answers may include that it is long and thin and runs north and south.

Week 22 Day 3 (page 122)

1. A river has water and food.
2. They need the river water to drink, grow crops, and travel on.

Week 22 Day 4 (page 123)

1. Amazon and Nile River
2. Answers may include that the river is wider or it begins at a lake.

Week 23 Day 1 (page 125)

1. Apple Street
2. homes
3. park

Week 23 Day 3 (page 127)

1. A city is more crowded. A suburb has mostly homes and a few businesses.
2. People can still work in the city. They can drive there.

Week 23 Day 4 (page 128)

1. Star Arcade
2. Gold Theater
3. Answers may include that it is easy to get to.
4. Students should tell how they used the chart.

Week 24 Day 2 (page 131)

Example: Go down Country Street. Turn left on Park Street. Turn right on City Street. Turn left on Farmer Street.

Week 24 Day 4 (page 133)

1. vegetables
2. nuts
3. 15 jars

Week 25 Day 1 (page 135)

Students should end up in the second desk in the right-hand column.

Week 25 Day 3 (page 137)

1. The students are wearing uniforms.
2. Kids get to school in different ways. There are all-girl and all-boy schools.

Week 25 Day 4 (page 138)

1. Lucia
2. Yuri
3. Students should explain how long their school days are.

Week 26 Day 1 (page 140)

1. There will be shade by the pool because there are trees next to the pool.
2. There are only a few places where flowers are shown on the map. I think it needs more.

Week 26 Day 3 (page 142)

1. Answers may include that it is well kept or that it has many plants.
2. A landscape architect plans where trees and flowers will be planted.

Week 26 Day 4 (page 143)

1. Answers may include that a city map shows more streets, and garden maps show a lot of plants.
2. City and garden maps both show where things are.
3. Students should add one detail to each part of the diagram.

ANSWER KEY *(cont.)*

Week 27 Day 1 (page 145)

1. a giraffe
2. five stops
3. north

Week 27 Day 2 (page 146)

east, north, west, west, east, east

Week 27 Day 3 (page 147)

1. People go on a safari to watch animals in their habitats.
2. Animals are endangered because people are taking over the land.

Week 27 Day 4 (page 148)

1. They live on only a quarter of what they used to have.
2. It shows that people have taken over the lions' land. The lions do not have as much land to live on.

Week 28 Day 1 (page 150)

1. Students should have labeled the water and land.
2. Answers may include: a globe and map both show where places are found; they both show land and water; and they both give the names of places.

Week 28 Day 3 (page 152)

1. A ushanka is a hat that has fur on the inside. It has flaps that covers a person's ears.
2. It looks foggy and hard to see. It might be snowing. There are not a lot of people on the street. They are probably inside staying warm.

Week 28 Day 4 (page 153)

1. August
2. January, February, March, December

Week 29 Day 1 (page 155)

1. 8 islands
2. Pacific Ocean
3. The islands are lined up like a chain.

Week 29 Day 3 (page 157)

1. A hot spot is where magma rises up through Earth's crust.
2. An island is land surrounded by water.

Week 30 Day 1 (page 160)

1. Kids could play on the playground, in the gym, and sometimes in the classrooms.
2. Students should circle the classrooms.
3. Students should draw kids on the playground.

Week 30 Day 3 (page 162)

1. Kids play tug of war and hand clapping games.
2. Example: yes. Everyone could make a long line. We would play it outside.

Week 30 Day 4 (page 163)

1. six votes
2. Answers may include that kickball is a game that a lot of kids could play together.

Week 31 Day 1 (page 165)

1. merry-go-round
2. tunnel, swings, and net
3. swing set

Week 31 Day 3 (page 167)

1. A geocache is the hidden object.
2. While they are hiding and looking for a geocache, they will see what is in their community.

Week 31 Day 4 (page 168)

1. Southeast
2. Northwest
3. Northeast

Week 32 Day 1 (page 170)

1. The mountain looks like a big triangle. It rises above the land around it.
2. A mountain rises up. A plain is flat and does not rise up.
3. Students should create symbols for mountains and plains.

Week 32 Day 3 (page 172)

1. A mountain is a tall and rocky part of land.
2. Hikers need a guide because climbing Mount Everest is hard. A guide will know how to do it safely.

ANSWER KEY *(cont.)*

Week 32 Day 4 (page 173)

1. Sandia Mountain must be a hard hike. It had the least amount of hikers make it to the top.
2. Mount Tanen must be easy to hike. It had the most amount of hikers make it to the top.
3. 750

Week 33 Day 1 (page 175)

1. Trail 2. It is the one that goes straight up the side of the mountain. The other trail is curvier.
2. Students should support their answers with reasons.

Week 33 Day 3 (page 177)

1. It takes longer because it is straight up the side of the mountain. It is a lot of work to climb a trail like that.
2. The shortest route is not always the fastest. The shortest route might be the hardest. People will go slow.

Week 33 Day 4 (page 178)

1. Rabbit Trail
2. Deer Trail
3. Pictures should show Rabbit Trail as the shortest and Deer Trail as the longest.

Week 34 Day 1 (page 180)

1. western
2. eastern
3. There are no volcanoes in the Great Plains. I know because there are no volcano symbols there.

Week 34 Day 2 (page 181)

1.

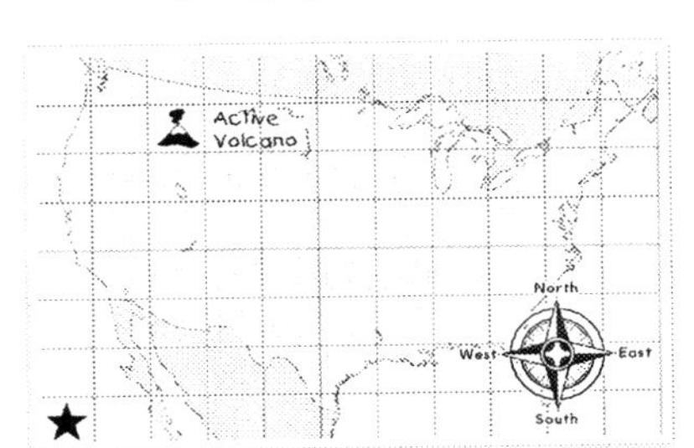

Week 34 Day 3 (page 182)

1. A volcano is shaped like a mountain. But not all mountains erupt lava.
2. The top of the volcano fell down in an earthquake.

Week 34 Day 4 (page 183)

1. mountain or hill
2. The vents should be labeled *magma*.
3. Lava should be drawn on the outside of the volcano.
4. An arrow should be drawn at the base of the ash cloud.

Week 35 Day 1 (page 185)

1. This cliff was formed near the ocean.
2. Students should shade the wall of the cliff.
3. Students should draw a green arrow pointing to the bottom of the cliff.
4. Students should draw a red arrow pointing to the top of the cliff.

Week 35 Day 3 (page 187)

1. On a foggy night, a boat might crash into the cliffs.
2. A lighthouse warns a boat to turn away from the cliff.

Week 35 Day 4 (page 188)

1. The entire landslide including the starting point and the debris at the bottom should be circled.
2. It could be dangerous to build a house on the edge of a cliff. If there was a landslide, the house might fall off.

Week 36 Day 1 (page 190)

1. Eagle Street
2. Dragon Flower Restaurant
3. Students should discuss their favorite restaurants.

Week 36 Day 3 (page 192)

1. People come from all over the world with their recipes.
2. There is food from all over the world.

Week 36 Day 4 (page 193)

1. Japanese
2. Booth 2
3. Japanese food, Hawai'ian food, Italian food, and American food.

Name: ______________________ Date: ______________

MAP SKILLS RUBRIC
DAYS 1 AND 2

Directions: Evaluate students' activity sheets from the first two weeks of instruction. Every five weeks after that, complete this rubric for students' Days 1 and 2 activity sheets. Only one rubric is needed per student. Their work over the five weeks can be evaluated together. Evaluate their work in each category by writing a score in each row. Then, add up their scores, and write the total on the line. Students may earn up to 5 points in each row and up to 15 points total.

Skill	5	3	1	Score
Identifying Map Features	Identifies and uses all map features, including the title, key, and compass rose.	Identifies and uses most map features.	Does not identify and use map features.	
Using Cardinal Directions	Uses cardinal directions to accurately locate places all or nearly all the time.	Uses cardinal directions to accurately locate places most of the time.	Does not use cardinal directions to accurately locate places.	
Interpreting Maps	Accurately interprets maps to answer questions all or nearly all the time.	Accurately interprets maps to answer questions most of the time.	Does not accurately interpret maps to answer questions.	

Total Points: ______________________

Name: ______________________________ Date: ______________

APPLYING INFORMATION AND DATA RUBRIC

DAYS 3 AND 4

Directions: Complete this rubric every five weeks to evaluate students' Day 3 and Day 4 activity sheets. Only one rubric is needed per student. Their work over the five weeks can be evaluated together. Evaluate their work in each category by writing a score in each row. Then, add up their scores, and write the total on the line. Students may earn up to 5 points in each row and up to 15 points total. **Note:** Weeks 1 and 2 are map skills only and will not be evaluated here.

Skill	5	3	1	Score
Interpreting Texts	Correctly interprets texts to answer questions all or nearly all the time.	Correctly interprets texts to answer questions most of the time.	Does not correctly interpret texts to answer questions.	
Interpreting Data	Correctly interprets data to answer questions all or nearly all the time.	Correctly interprets data to answer questions most of the time.	Does not correctly interpret data to answer questions.	
Applying Information	Applies new information and data to known information about places all or nearly all the time.	Applies new information and data to known information about places most of the time.	Does not apply new information and data to known information about places.	

Total Points: ______________________________

Name: ______________________ Date: ____________

MAKING CONNECTIONS RUBRIC
DAY 5

Directions: Complete this rubric every five weeks to evaluate students' Day 5 activity sheets. Only one rubric is needed per student. Their work over the five weeks can be evaluated together. Evaluate their work in each category by writing a score in each row. Then, add up their scores, and write the total on the line. Students may earn up to 5 points in each row and up to 15 points total. **Note:** Weeks 1 and 2 are map skills only and will not be evaluated here.

Skill	5	3	1	Score
Comparing One's Community	Makes meaningful comparisons of one's own home or community to others all or nearly all the time.	Makes meaningful comparisons of one's own home or community to others most of the time.	Does not make meaningful comparisons of one's own home or community to others.	
Comparing One's Life	Makes meaningful comparisons of one's daily life to those in other locations all or nearly all the time.	Makes meaningful comparisons of one's daily life to those in other locations most of the time.	Does not makes meaningful comparisons of one's daily life to those in other locations.	
Making Connections	Uses information about other locations to make meaningful connections about life there all or nearly all the time.	Uses information about other locations to make meaningful connections about life there most of the time.	Does not use information about other locations to make meaningful connections about life there.	

Total Points: ______________________

MAP SKILLS ANALYSIS

Directions: Record each student's rubric scores (page 202) in the appropriate columns. Add the totals, and record the sums in the Total Scores column. Record the average class score in the last row. You can view: (1) which students are not understanding map skills and (2) how students progress throughout the school year.

Student Name	Week 2	Week 7	Week 12	Week 17	Week 22	Week 27	Week 32	Week 36	Total Scores
Average Classroom Score									

APPLYING INFORMATION AND DATA ANALYSIS

Directions: Record each student's rubric scores (page 203) in the appropriate columns. Add the totals, and record the sums in the Total Scores column. Record the average class score in the last row. You can view: (1) which students are not understanding how to analyze information and data and (2) how students progress throughout the school year.

Student Name	Week 7	Week 12	Week 17	Week 22	Week 27	Week 32	Week 36	Total Scores
Average Classroom Score								

MAKING CONNECTIONS ANALYSIS

Directions: Record each student's rubric scores (page 204) in the appropriate columns. Add the totals, and record the sums in the Total Scores column. Record the average class score in the last row. You can view: (1) which students are not understanding how to make connections to geography and (2) how students progress throughout the school year.

Student Name	Week 7	Week 12	Week 17	Week 22	Week 27	Week 32	Week 36	Total Scores
Average Classroom Score								

DIGITAL RESOURCES

To access the digital resources, go to this website and enter the following code: 22662988.
www.teachercreatedmaterials.com/administrators/download-files/

Rubrics

Resource	Filename
Map Skills Rubric	skillsrubric.pdf
Applying Information and Data Rubric	datarubric.pdf
Making Connections Rubric	connectrubric.pdf

Item Analysis Sheets

Resource	Filename
Map Skills Analysis	skillsanalysis.pdf skillsanalysis.docx skillsanalysis.xlsx
Applying Information and Data Analysis	dataanalysis.pdf dataanalysis.docx dataanalysis.xlsx
Making Connections Analysis	connectanalysis.pdf connectanalysis.docx connectanalysis.xlsx

Standards

Resource	Filename
Standards Charts	standards.pdf